Books by

COMMANDER ELLSBERG

ON THE BOTTOM

PIGBOATS

S-54

HELL ON ICE

THIRTY FATHOMS DEEP

OCEAN GOLD

SPANISH INGOTS

MEN UNDER
THE SEA

COMMANDER EDWARD ELLSBERG

MEN UNDER THE SEA

by

COMMANDER
EDWARD ELLSBERG

GREENWOOD PRESS, PUBLISHERS
WESTPORT, CONNECTICUT

Library of Congress Cataloging in Publication Data

Ellsberg, Edward, 1891–
   Men under the sea.

     Reprint. Originally published: New York : Dodd,
   Mead, 1939.
       1. Diving, Submarine--History.   2. Salvage--History.
   3. Ellsberg, Edward, 1891–   .   I. Title.
   VM977.E44   1981   627'.72   81-6869
   ISBN 0-313-23030-7 (lib. bdg.)   AACR2

Reprinted with the permission of Dodd, Mead & Company.

Reprinted in 1981 by Greenwood Press
A division of Congressional Information Service, Inc.
88 Post Road West, Westport, Connecticut 06881

Printed in the United States of America

10 9 8 7 6 5 4 3 2 1

TO

LIEUTENANT OLIVER NAQUIN, U.S.N.,

AND

THE CREW, LIVING AND DEAD,

OF

U.S.S. "SQUALUS"

HEROES ALL

PREFACE

THE world beneath the sea has always had a peculiar fascination for us mortals, though unequipped ourselves by nature with fins, tails, and gills to explore its watery depths as may the fishes. But our lack of natural equipment has never prevented our imaginations from running riot with the ocean's wonders and its treasures, and from ancient days to the present the literature of the deep sea, whether presented as fact or fiction, has mainly been notable for the fantastic tales which adorn it and the unreal terrors and legendary treasures with which authors have strewn the ocean floor.

MEN UNDER THE SEA is the story of real men of several nations and of various centuries who in different parts of the ocean and for diverse reasons have actually battled with the deep sea, sometimes for treasure, sometimes to save others, sometimes to save themselves. Here they are, whether trapped in smashed submarines, dressed in rubber diving suits, cased in ponderous metal armor, or as naked as nature brought them into the world, face to face with the real terrors of the deep. Sometimes they lived and won, sometimes they died— in either case their adventures under the sea need garnishing neither with fabulous monsters nor mythical treasures to make them as gripping to us today as anything evolved in the fertile imaginations of Victor Hugo or Jules Verne or of their more modern imitators.

In some of the struggles related here I took part. In most of the others I have had personal contact with some of the participants and with what took place, and can vouch for the facts, or (as will be noted occasionally) for the lack of them in the instances cited where the public has been gulled by self-styled salvage experts(?) and their glowing tales of wealth beneath the seas to be recovered by their particular diving gadgets.

For details in connection with the *Laurentic*, I am indebted to Captain G. C. C. Damant, Royal Navy (now retired), who commanded throughout in the unprecedented recovery of $44,000,000 in gold from that wartime victim of German submarine activities.

For data on the Davis "lung," for much of the history of diving, and for permission to use certain illustrations from his book DEEP DIVING, thanks are due to Sir Robert H. Davis, manager of Siebe, Gorman & Co., of London, pioneers for over a century in diving developments and equipment.

The illustrations and much of the data on the use of the underwater torch were furnished through the courtesy of Mr. Charles Kandel, manager of the Craftsweld Equipment Corporation of New York.

To Mr. William Wallace Wotherspoon, salvage master on many wrecks in the United States and Canada, I owe my information on the *Empress of Ireland*.

To the late Lieutenant Alberto Cuniberti, Italian Royal Navy, lamented martyr to the advancement of safety beneath the seas, I owe a debt which can now never be paid for much information on deep diving in armored rigs, and in particular for information on the salvage of the *Egypt's* treasure.

Special thanks are due to Commendatore Giovanni Quaglia, president of Sorima of Genoa, Italy, and to Signor Galeazzo Manzi-Fè de Riseis, manager of Sorima, for many courtesies while in Genoa, for detailed information, and for their privately printed account (on which my story is mainly based) of Sorima's successful recovery of a huge amount of bullion from the *Egypt's* strong room in by far the deepest salvage job ever attempted by men under the sea.

In this connection, to those interested in a very stirring and completely detailed eye-witness account of the *Egypt* salvage, I can recommend the two-volume story, SEVENTY FATHOMS DEEP and THE EGYPT'S GOLD, by David Scott, correspondent of the London *Times*, who was present during much of the work on the *Egypt*.

To Lieutenant Commander C. B. Momsen, U.S.N., I am grateful for data furnished on the Momsen "lung" now used for submarine escape in the United States Navy, and for the account of our Navy's recent experiments in deep diving with helium. To Lieutenants A. R. Behnke and O. D. Yarbrough, Medical Corps, U.S.N., I am obliged for most of the facts on the physiology of "lung" training and of the physiological effects of helium in deep diving.

Lastly to Commander Henry Hartley, U.S.N., late skipper of the U.S.S. *Falcon*, and to my old shipmates on the *Falcon*, who at the risk of their lives unwittingly furnished many of the adventures here related, I owe a deep debt of gratitude. Especially is my appreciation here extended to Chief Bosun's Mate Bill Carr, who in spite of my one lamentable failure to coöperate with him, finally and deservedly won the Navy Cross for heroism on the *S-4*.

EDWARD ELLSBERG.

ILLUSTRATIONS

MEN UNDER
THE SEA

CHAPTER I

SHROUDED in a web of frayed hawsers and dripping air hoses, a battered submarine, with a ragged gash laying open her port side from deck to keel, rested in the dry-dock. With her diving fins cocked drunkenly in opposite planes, her conning tower half smashed in, her rudder jammed hard astarboard, and a trickle of mud and water oozing from her stern torpedo tube, that submarine was a dismal sight. Around her, crazily floating half awash in the nearly unwatered drydock, were the eight huge cylindrical pontoons which had floated that wreck in from her ocean bed at the bottom of the cold Atlantic off Block Island, 150 miles away.

Looking down on that submarine from the towering sides of the drydock were the divers who had lifted her and brought her in, their incredulous eyes hardly able yet to believe that actually they saw her there, the submarine S-51, for whose hulk over nine seemingly endless months they had battled the fierce Atlantic—that at last, safely locked in behind the caisson of that drydock holding back the sea, they had that ship where she could not again get away from them and sickeningly slip from their sight back to the ocean depths.

They were a wan group of men, those staring divers —weather-beaten, with cracked lips, seamed faces, sunken eyes, and lean bodies from which had been burned every ounce of fat by long hours of breathing excessive oxygen in heavily compressed air forced down to them as they struggled on the ocean floor. For the hundredth time they leaned again over the drydock rails, gazing unbelievingly at the ship they had salvaged. Hopeless their task had seemed when, 22 fathoms down at the bottom of the icy ocean, they first had pitted their puny bodies against the powers of the sea, to lift that 1,200-ton wreck to the surface. Even more hopeless the task had seemed after months of fruitless struggling far offshore, when—in the minds of the divers groping in mud, in darkness, and in frigid cold—the ocean began to take on a definite personality, that of a malignant demon with superhuman cunning and unearthly strength fighting to hold from them what it had claimed as its own. Relentless was the grip, formidable were the weapons of the sea—on the surface, fierce gales to batter and to scatter the salvage ships; on the bottom, icy cold water to freeze a man to the very marrow of his bones, the dark solitude of the weird depths to drive cold fear into a man's heart as he struggled alone in the mud and the blackness enfolding that wrecked submarine, a cold fear to paralyze the heart even more than the chilling water ever could paralyze the body, and last and worst of all the terrible pressure of the deep sea enveloping everything, ready instantly to crush a man into jelly under a load of 60 tons should he by any mischance lose the air pressure inflating his suit.

Well, all that was over now and they had won. Singly and in knots of two or three, the divers gaped down into the chasm of that drydock at the ship they had lifted—

IN THE DRY DOCK, SAFE ON THE KEEL BLOCKS BEHIND THE MASSIVE
CAISSON, RESTED THE "S-51"

slight Francis Smith, who once, buried in a tunnel cave-in beneath that boat, had by a miracle dug his way out; big Jim Frazer, who would never dive again, his heart dilated from the superhuman strain of dragging a heavy hatch along the deck of that sunken sub; lanky Fred Michels, who had narrowly escaped being crushed between a swaying pontoon and the conning tower; profane Tug Wilson and silent Joe Eiben, who in cumbrous diving rigs had, like eels, snaked their way through the narrow doors inside that flooded submarine when each door meant a desperate gamble with death on getting back; slow-spoken John Kelley, who with a flaming torch on the bottom of the sea had sliced through obstructing steel like butter; Tom Eadie, ace of divers, who nevertheless had almost drowned in his own diving rig when a strange accident in the depths had ripped his canvas suit wide open; and red-faced Bill Carr, on deck a belligerent bosun's mate, on the bottom as steady-going a diver as ever I worked with.

It was over, certain enough. In the drydock, safe on the keel blocks behind the massive caisson, rested the S-51. The salvage squadron could disband now—the Falcon, our diving ship, to rejoin the submarine flotilla at Panama, the other vessels, their fleet stations, and the divers (torpedomen and gunners' mates, most of them), after brief leaves ashore, to scatter again to the various ships from which they had hastily been gathered when the S-51 had been rammed and sunk. For the men who fought through that heart-wrenching submarine campaign, nothing remained except to pack bags and hammocks and depart. The last order in the salvage squadron had shortly before been posted where the men returning aboard from the drydock would see it—the list of honors recommended for the men who had raised the S-5 1—Navy Crosses to six of the divers who had heroically distinguished themselves in extraordinary circumstances, promotions and letters of commendation to certain others whose services on the bottom had been only routine (if such a word can be used with regard to as perilous a job as diving).

In a tiny cabin on the *Falcon*, the scene of many a heartache while working over the S-5 1, I listlessly gathered up my few belongings preparatory to going ashore myself. I was through, also, as Salvage Officer. Slowly I gathered up my blueprints, my instruments, my records, subconsciously as little able as any of my divers to realize that our seemingly endless task was over, that I should not again have to turn to over those designs of the submarine to figure out a way of untangling another knot that the sea had unexpectedly tied in our plans; that I should not again have to drape myself in that 200 pounds of lead and copper and wet canvas that made up a diving rig, and drop to the bottom of the cold sea to struggle on that submarine, while I mapped out their work, with the same dangers that my divers faced.

Never again should I go through that. Once in a lifetime was one time too many. Sixteen years of the Navy topped off with those nine months battling the depths off Block Island was enough of a naval career for me, and when my report was in the hands of the Navy De-

partment, I was through with the sea forever. I pulled
my slide rule from a cubbyhole in the desk and tossed
it in on top of the *S-51's* blueprints, wondering whether,
even as a civilian, I should ever get over that leaden
feeling in the stomach which griped me every time a
diver slipped over the *Falcon's* side to disappear in a
swirl of bubbles as his copper helmet vanished beneath
the sea, or that mental agony as on deck I listened on
his telephone, the while he groped in the mud and
the wreckage on the bottom, over what the treacher-
ous sea was likely to do to him?

I began rolling up the blueprints. Some day, perhaps,
when I had got back on myself the flesh that excessive
oxygen under pressure had burned off my bones down
there on the ocean floor, I supposed I might get over
the sickening dread of momentarily expecting that, in
spite of all my planning and my care, the next instant
something would go wrong and that someone for whom
I was responsible would suddenly be in deadly peril,
fighting in solitude and in darkness for his life against
the—

BANG!

A vicious rap on the cabin door. I started involuntarily, dropped my roll of plans. Before I could say
anything, the door flew open and there, framed in it
against a background of signal flags and rigging outside, stood Bosun's Mate Bill Carr, his bellicose face a
fiery red, his blue eyes flashing, and waving a paper
clutched in his brawny fist. I looked at him in astonishment. An unceremonious entrance into an officer's
cabin, to say the least, but Bill Carr apparently wasn't

standing on ceremony that day.

"Say, Mr. Ellsberg, look at that!"

For an instant Carr thrust under my nose the paper he was clutching. It was the order containing the list of divers' rewards, evidently torn away from the bulletin board at the gangway.

"This order's all wet, Mr. Ellsberg!" bellowed Carr. "Here it says that Jim Frazer an' Tom Eadie an' John Kelley an' Francis Smith an' Tug Wilson an' Joe Eiben have all been recommended fer Navy Crosses fer their work on that sub, an' all I'm down fer is a letter o' commendation fer what I did! A letter o' commendation! It ain't right, commander! I've earned a Navy Cross as much as anybody, an' I want it!"

I looked at Carr. His stocky figure, tense with anger, filled every inch of space in that tiny cabin between bunk and desk, and his blazing red face, so close to mine, left no doubt that he was in deadly earnest, that he felt badly cheated. I sympathized with Carr, but he was wrong and there was nothing for it but to convince him so.

"Sorry, Carr, but haven't you overlooked something?" I asked as mildly as possible. "Don't forget that, aside from that letter of commendation that's griping you so much, you're down on that order for promotion to chief bosun's mate as well."

"To hell with that promotion!" barked Carr. "What's it amount to in my case? I'd 've made a chief's rate before this cruise's over even if I'd never seen a diving rig! But when'll I ever git another chanst at a Navy Cross? Answer me that!"

I eyed my bosun's mate curiously. Ordinarily such insubordinate language warranted a call for a master-at-arms and a transference of further discussion to a deck court, but this wasn't an ordinary occasion and Carr's previous services certainly entitled him to a hearing, however informal he chose to make it. Somewhat perplexed, I undertook to soothe his ruffled feelings.

"Look here, Carr, you're getting promoted from first class petty officer to chief, right now. And, in spite of what you think of yourself, you might never make a chief's rate this cruise or any cruise. And on top of that, you'll get a letter from the Secretary of the Navy specially commending you for your services on the S-51. That's a lot for what you've done."

"Fer what I've done?" Carr bristled perceptibly. "Commander, I made more dives in salvaging that sub than any other gob in the outfit, an' your own records'll show it. You look. So if anybody rates a Navy Cross fer that job, I'm the lad that oughta be gittin' it!"

To a degree, Carr had me there. A horse for work and a bull in physique, Carr *had* made the most dives— I knew that without checking the records. When other divers were tied up with "the bends," knocked out by an accident, or from sheer exhaustion unable to don a rig and go overboard, stocky Carr had never missed a turn. But in the Navy, nobody gets medals just for routine work. I wondered if I could make my bosun's mate understand that.

"Yes, Bill, I know that. You made more dives than anybody. I'm not disputing it with you. That's already been carefully considered. Still, you're out of luck just the same. There were more than twenty divers, including you, on that job. But only six of 'em, the boys you mentioned, are getting Navy Crosses. And why only six? Because they did something especially heroic on the bottom. You didn't. I'm not saying you couldn't have, Bill, any less than those who did, but if it was your hard luck that things broke down on the bottom so they got a chance to be heroes and you didn't, I can't help it. So you'd better take your promotion and that letter of commendation and be happy. It's a lot more than some of your shipmates are getting. Come on, Carr; say you're satisfied."

But Carr was obdurate. Satisfied? It was obvious that he wasn't. However, my logic seemed to have had some effect, for he dropped his bluster, took a different tack, and tried wheedling me into compliance. The truculence in his tone faded out, his red face relaxed into a broad grin, and his blue eyes took on a friendly twinkle I had never seen there before.

"Aw, now, commander, be a good fellow. Gimme that Navy Cross anyhow. Sure an' you know I've earned it! An' what good's a letter o' commendation to me anyway? I reads it once an' stows it away in my diddy box an' nobody ever knows I got it. But a Navy Cross is different. There's Jim an' Tom an' Tug an' Joe an' those other lads that's gittin' 'em can pin their Navy Crosses on their coats every time they makes a liberty, an' the girls ashore all knows they're heroes! How about me? Can I pin that letter o' commendation on my chest when I goes ashore so the girls'll know

I'm a hero too? Like hell I can! Come on, now! Be a
sport, commander, an' gimme that Navy Cross! It'll
look grand on that new chief's uniform you're pro-
motin' me to!"

But by all Carr's pleas, as by his bluster, I was un-
moved. Fate had never placed him on the *S-51* in circumstances of extraordinary peril where he could distinguish himself by heroism above and beyond the line
of duty, and I could not recommend him for a Navy
Cross. Again and again I patiently reiterated that, and
finally, having worn him down somewhat, I managed
to ease the still grumbling bosun's mate out of the little cabin and wearily to complete my packing.

The salvage squadron disbanded that day. With a
heavy heart I said good-by to the divers on the *Falcon*,
my companions on the bottom through nine terrible
months. Never again would I see them, as I was leaving the Service and they were going back to ships scattered over the seven seas.

Once ashore, I toiled in the New York Navy Yard
over my salvage report until that, laden down with intricate computations and detailed plans showing how
we had raised the *S-51*, was on its way to Washington.
Then I doffed my uniform for the last time, slid into
civilian clothes, and, after sixteen years in the Navy,
was once again just a civilian.

A year and a half went by uneventfully and Bill
Carr and his troubles, submarines and salvage, had
gradually faded out of my quiet suburban life in a lit-

tle town in New Jersey, when, one cold Sunday morning in December, 1927, a week before Christmas, I opened my front door to reach for the Sunday paper lying on the steps, only to be frozen into immobility by a flaring headline screaming at me in large type:

SUBMARINE S-4 SUNK!

                              FORTY MEN TRAPPED!

CHAPTER II

ABOUT noon on December 17, 1927, the U.S.S. S-4 had proceeded from inside Provincetown Harbor to the deep water trial course off the tip of Cape Cod for submerged standardization trials. For some months before, at her home Navy Yard, Portsmouth, New Hampshire, the submarine had been undergoing repairs and refitting; now she was to be run submerged under practically laboratory conditions to determine the effects on her submerged speed and maneuvering qualities.

From the near-by Coast Guard station at Wood End on the sandy tip of Cape Cod, overlooking the trial course and not half a mile away, were flying signals warning of an approaching northwest storm--for that vicinity the worst possible direction, as the wind would have a free sweep down the coast and across all of Massachusetts Bay before striking the unsheltered trial course.

The day was cold, the sea already rising, with whitecaps everywhere, and a force 4 wind whipping up a stiff chop over the whole bay, as the S-4, leaving her tender inside Provincetown Harbor, moved slowly out of the protected waters to the open bay, under the direction of Lieutenant Commander R. K. Jones, her captain for over two years.

The S-4, designed and built by the Navy, was a

double-hulled submarine 231 feet long, 22 feet in the
beam, and of 800 tons surface displacement. Of this
special class, the S-type, built mainly during the World
War and considered generally a very satisfactory size
for all-around service, the Navy had about 50 boats. On
the surface, the *S-4* was driven by two 8-cylinder Diesel
engines; submerged, by two powerful electric motors
from massive storage batteries.

For safety in case of accident, four heavily reinforced
transverse steel bulkheads divided the *S-4* into five main
watertight compartments. These were, in order, starting from the bow, the torpedo room, the battery room,
the control room, the engine room, and the motor room.
Of these compartments, the battery room, owing to
the space required to house the storage cells which fed
the submerged propelling motors, was by far the largest.
It extended practically from the conning tower amidships some 51 feet forward through the widest part of
the vessel and provided incidentally, in the space over
the battery cell storage, the living and sleeping quarters
(much confined of course) for the entire crew, both
officers and men, except a few torpedomen whose
berths were slung over the torpedo storage forward.

On this particular December day, in addition to her
regular crew of 4 officers and 34 men, the *S-4* carried,
to observe the trials as representative of the Navy Trial
Board in Washington, Lieutenant Commander Callaway and his civilian assistant, Mr. Charles Ford, making a total of 40 aboard.

Meanwhile (it being still in the heyday of Prohibition), at 9 A.M. that morning, the Coast Guard De-

stroyer *Paulding*, one of a fleet of 25 such vessels transferred from the Navy to the Coast Guard mainly for
the prevention of rum-running, steamed out of Boston
Harbor for a sweep at high speed through Massachusetts Bay and outside Cape Cod. It was assumed that,
with the approach of the holidays, increased activity
of rum-runners known to be operating off that coast
could be anticipated, and the presence in those waters
of at least one "notorious offender" was suspected. At
a speed of 18 knots, the *Paulding* cleared Boston Harbor and headed out to sea. On her bridge, aside from
Lieutenant Commander John S. Baylis, her captain and
and an officer of long experience in the Coast Guard,
there were on watch a junior officer and three of the
crew, all on the alert to scan and identify every vessel
which hove in sight.

The *Paulding* had for three hours been speeding
through the rising storm when about noon the *S-4*
slowly nosed out of Provincetown, her interior
crowded as always, especially crowded this day by the
added equipment in the cramped control room for accurately registering the speed of the shafts during the
trials. As she moved away from the tender, with the
interior of the boat throbbing to the vibrations of the
Diesels, the crew taking their stations shivered from the
chill inside the hull as the blasts of cold December air
swept through the control room and aft to be sucked
into the intakes of the Diesels. Forward in the torpedo
room were Lieutenant Fitch and his torpedomen, little
concerned this day with the engine trials. Amidships in
the control room and aft in the engine and motor rooms

were the rest of the crew—the captain; Lieutenant McGinley, the navigator; Lieutenant Weller, the chief engineer; and the Trial Board representatives.

Surrounding them in the control room on all sides was machinery. On the starboard side was the switchboard, a glittering array of electrical switches of all sizes for the complicated machinery of the boat. On the centerline were the periscopes, glistening steel tubes with their eyepieces hidden in housing wells below the deck. And to port was a dizzying conglomeration of pressure gauges, air valves, flood valves, drain valves, diving wheels, depth gauges, periscope motors, and all the intricate mechanism for controlling the submerged operation of the vessel. Here also was placed the special counter gear which Callaway and Ford had brought for calibrating the speed, and running aft from these instruments through the opened after watertight door of the control room were strung, temporarily, electric cables to the propeller shafts in the motor room.

A few miles out of Provincetown, the S-4 approached the trial course, marked approximately by two white can buoys half a mile offshore, and more exactly delineated at each end of the "measured mile" by a range mark of two poles set up perpendicular to the course. Beneath the sea between these two range marks, the S-4 was carefully to calibrate her speed and power.

Inside the S-4 the raucous note of electric horns cut through the clatter of the engines. The diving signal. All over the boat men sprang to diving stations. Another signal and the conning tower hatch slammed to, ventilation valves were closed outboard, Diesels hastily shut down and unclutched, kingston valves jerked open to flood the ballast tanks, and the *S-4*, driving ahead on her electric motors, planed smoothly down to periscope depth and commenced her trials.

Gone now were the roar and clatter of the Diesels, the slap of the waves against the rounded hull, the rolling of the ship in the seaway. Except for the clicking of the revolution counters, the slight whir of ventilation fans exhausting and circulating the battery gases to dissipate them, and the nearly imperceptible hum of the main propelling motors, silence filled the boat as, with her deck 20 feet below the surface, she swam down the course, passed the first buoy close aboard and headed southwest with only her two periscopes showing a few feet above the surface, one periscope with its lens fixed on the range marks ashore, the other with its solitary eye sweeping the horizon as lookout for passing vessels.

From the Coast Guard observation tower at Wood End, half a mile north of the course, during the next three hours, Surfman Frank Simonds, lookout on watch, saw off and on the periscopes of the *S-4* swinging back and forth over the measured mile between the two can buoys. Neither her hull nor conning tower ever showed above surface, just a few feet of the periscopes could be seen cutting the rough water with a tiny "feather" or wake of spray following the thin periscope fingers as they sliced through the water.

At 2:46 P.M. the *Paulding* concluded her patrol in the open sea and, on a westerly course, headed in for a sweep past Provincetown Harbor and through Cape

Cod Bay. Outside nothing of importance to her mission had been sighted. A few minutes later, Race Point Light was rounded and the destroyer started to skirt the fishhook tip of Cape Cod, heading southeast on a course which would take her well clear of the ranges off Wood End. A fishing vessel was swiftly overhauled, identified as the *William Langtry* of Boston, and passed without further notice. Wood End Light was drawing abeam; storm signals were flying there. The quartermaster swung his glass to read the flags, and at this time, 3:33 P.M., having followed the southeast course for over three miles, the *Paulding* passed a sea buoy off Wood End and abruptly changed course to port, heading 94° (practically east); for the first time pointing directly for the trial course which, to all eyes on the *Paulding*, seemed clear of shipping.

In the Coast Guard station ashore, Boatswain Gracie, in charge, climbed the tower, popped up through a trap door into the observation room.

"What's doing, Frank?" he asked of his lookout.

"Not much, sir," replied Simonds. "I've seen a submarine operating under the beach."

Gracie took the telescope, focused it on the *Paulding*, and, noting her easterly course, became suddenly alarmed.

"Frank, I wonder where that submarine is now? Have you seen her?"

"No, sir; not lately."

Hurriedly Gracie swung his telescope to the southeast on the can buoy marking the near end of the measured mile. There, centered in his glass, headed toward the destroyer, was the flash of the periscopes, a streak of spray flying in air! For a second only he watched, then lowered the telescope, looked again at the *Paulding*, and shouted:

"My God, Frank, there's going to be a collision!"

Gracie dropped his telescope, raced down the tower to get his lifeboat underway.

On the *Paulding*, the buoy off Wood End having been rounded, the course was set east to clear on the port hand the next can buoy (the one marking the end of the course) about a mile ahead. The captain dropped back to the chart house in the rear of the bridge to study the chart. The junior officer of the deck, in the starboard wing of the bridge, picked up in his glass, several miles off on the starboard bow and headed for them, the Nantucket Lightship, evidently off her station, and studied her to make sure before entering the fact in the log.

On the port side of the pilot house, the quartermaster, searching with his glasses for storm signals in Provincetown Harbor itself, looking off to port, saw suddenly about one point on the port bow and not over 200 feet away, two periscopes. Simultaneously, Ensign Phanenmiller, officer of the deck, also picked them up and shouted:

"Hard astern! Full right rudder!"

As the *Paulding* during the next ten seconds strove desperately to reverse engines and swing to starboard, before the horrified eyes of her officers the two periscopes lifted from the water, half the conning tower of a submarine broke surface right under their port bow, and then came a terrific crash as the hurtling destroyer struck! The *Paulding's* bow rose as she drove on. For an instant the tapered stern of a submarine lifted drunkenly above the surface and drifted down the port side, visible a moment abreast the destroyer's smokestacks, then vanished. Except for bubbles and a little oil slick, nothing again showed on the surface as the quivering *Paulding* came to a stop, frantically lowered a lifeboat, dropped a buoy to mark the spot, and hastily took cross bearings of the lights ashore to determine her position.

The *S-4*, which had been planing upward to surface, with her periscopes already half housed on their way down and useless for observation, reeled from the blow as if hit by a giant sledge, rolled heavily to port, and then, with her battery room torn open, began to sink bow first. A torrent of water poured through to flood the battery compartment. In the torpedo room forward, toward which the water ran first, Lieutenant Fitch and his five men found their path to the conning tower amidships and what chance of escape it offered, blocked off by that Niagara cascading into the room between them and the sole escape lock in the boat. With quick death staring them in the face, they slammed the torpedo room door shut against the water already pouring through it, hastily jammed down the dogs, and sealed themselves up, six men altogether, in the torpedo room.

In the control room, crowded with men and officers, conditions were worse. The forward periscope, hastily housed when collision was inevitable, came down with its training handles still rigged out and jammed itself in its housing well, while the hoisting wires, still slacking off as the motor continued to spin round, spread themselves in snaky coils helterskelter over the deck to tangle the feet of men, still reeling from the shock of collision, trying automatically to get back to their stations.

"BLOW ALL BALLASTS!"

At the blowing manifolds, swift fingers traveled over the valves, frantically opening compressed air lines from high pressure air banks No. 1 and No. 2 to every main ballast tank in the boat—forward, amidships, aft—hurriedly to force out the water there, to lighten up, to float the boat to the surface before she went too deep. But the forward ballast tank of the S-4 was now torn wide open. Uselessly the precious air whistling through the blowing lines escaped to the sea without displacing any ballast, and, with fresh tons of water rushing in each moment, the S-4 only accelerated her downward plunge!

Water rising in the battery room! Someone leaped forward along the narrow passage in the control room to close the forward door. There was room only for one man to work there. But Fitch and his five torpedomen were forward. What of them? Where was the damage— in the battery room? In the torpedo room? Perhaps in both? In the control room nobody knew, nobody could know. For an instant perhaps the door was held open, but the missing men did not come aft. And then, in the face of the rising water, pouring through faster and faster as the boat sank and sea pressure increased, rising now to flood over the high sill, the steel door to the battery room was swung shut, a few dogs turned down to hold it. No more were necessary, for the water pressing that door against its seat would soon enough jam it tight.

Bow first, at a sickening angle, the S-4 went down. For the moment, the sea had been shut out; the men in the control room struggled to free themselves of the tangling coils of periscope wire and man all the controls. With diving rudders at "RISE," air roaring through to blow ballasts, they had done everything possible in the emergency to start the boat up. Agonized eyes watched the depth gauge dials, but inexorably the needles went up the scales, continuously registering a greater and greater depth. 80 feet—90—100—

CRASH!

Again the boat reeled. Bow first she had struck bottom hard, plowed heavily along a few feet in the mud, then leveled off on an even keel.

The S-4 was on the bottom in 110 feet of water. To Lieutenant Commander Jones, to his men there in the control room, must have come an instant of hope. Things were not so bad. The sea was sealed out forward, most of the crew were safe aft, best of all they were in full possession of the control room with all its machinery, its controls, and the precious air still left in banks No. 3 and No. 4. By themselves they might raise at least the undamaged stern of the boat and escape that way. From even deeper water off the Delaware Capes, the crew of their sunken sister, the *S-5*, had done that very thing some six years before. So might they.

And then came disaster.

From overhead in the control room itself, a geyser of water burst suddenly forth, spraying directly on the live electric contacts of the switchboard to starboard! Wild eyes swept upward, seeking the source. Plain enough. A thin sheetmetal ventilation duct overhead, intended to carry the exhaust gases from the storage batteries forward to the engine suctions aft, had burst wide open, rupturing, of all places, only in front of that switchboard, and now was deluging it with salt water!

"CLOSE THAT FORWARD VENTILATION VALVE!"

In the confined forward passage, where that ventilation duct came into the compartment through the bulkhead just over the door, was a lever-operated quick-closing valve, intended when necessary to seal off that duct watertight at the bulkhead. The man at the forward door, who so far had turned down only three of the securing dogs on the door, left off clamping down dogs, reached overhead for the valve lever, hastily pushed it closed.

But to the dismay of all those abaft him in the control room, the deadly spray of water only diminished somewhat but did not stop! On the switchboard, circuits began to short, vivid flashes leaped like lightning between the contacts, burned the man at the board. Forward, in the restricted space before the valve, the solitary man who could get in there to work struggled desperately to seal it tight, while astern of him, amidst the tangles of periscope wire, his shipmates fought even more desperately to protect the switchboard from the water spraying on the contacts, to prevent further damage. Obtained in that mass of machinery from God knows where, a screen of canvas and cloth was hastily improvised and draped across the switchboard to shield it from the flood.

But through the ruptured duct the deadly stream of water under high pressure kept pouring in; the bulkhead valve would not swing home. Unknown to the men in the control room, beyond the bulkhead the ventilation duct in the battery room had collapsed under the sudden impact of the sea pressure and torn away from the valve body on the battery room side. The rising water in the battery room had floated up on its surface a green baize curtain draping the door of the captain's stateroom just forward of the control room bulkhead. When the water reached the valve which the collapsing duct had just exposed, it poured aft through the opening, picked up the curtain in the rushing stream, and washed it into the valve body, effectually preventing the valve disks from seating.

Around that valve the major battle inside the S-4 was fought. Temporarily the switchboard was shielded, the short circuits stopped, but unless that valve was soon seated tight, the control room and all it meant to 34 desperate men was lost. In the narrow passage leading to the door only one man at a time could get to the valve. In the restricted space overhead, between partitions and piping, fumbling fingers and straining arms fought against despair to jam that valve lever home.

They failed. That harmless green baize drapery, clinging like a leech to the inside of that valve, was more deadly now to the crew of that submarine than depth bombs and TNT. Against those unseen folds fouling the valve seat, the men of the *S-4* fought in vain as the sea poured through and steadily the water rose on the deck in the control room, flooding the pumps, lapping upward toward the switches, rising inexorably toward the level of the high sill on the after door, when it would cascade over into the engine room.

Beaten by that valve which would not close, the struggling officers and men were forced to abandon the control room with its escape lock, its compressed air, its controls, its chances of expelling fuel oil and enough ballast water from the undamaged after tanks at least, to float up the stern—to abandon everything that to a submarine sailor means anything, and flee helplessly before the flood into the engine room while yet there was a chance to flee.

The control room was abandoned; with its loss went all hope of doing anything for themselves. The last man squeezed through, the engine room door swung shut, the 16 heavy dogs on it were speedily twisted home. Soberly from its after side, 34 trapped men, helpless now to help themselves except to hold the sea out of their prison, regarded the door. That door swung closed against the after side of the bulkhead. The sea pressure on the other side would tend to spring the door and its rubber gasket away from its frame against which the dogs were clamping it. Would the dogs hold? Were they safe here at least from drowning?

Swiftly on the other side of the bulkhead the control room flooded, the pressure built up against that door. Then before the horrified eyes of the crew, as the full sea pressure came at last against the bulkhead, on one side the door gave a little, and under heavy pressure a flat sheet of water sprayed on through into their last refuge!

Once more the battle to hold out the sea commenced. Overwrought men sledged down the dogs over the tapered brass wedges on the door till the dogs brought up against their stops and would go no farther. But still some water gushed on through. Again the clang of the hammers rang out through the sea in a frenzied attempt, stops or no stops, to tighten the dogs still more, only to have five brass wedges shear off under the hammer blows, releasing completely all hold on those dogs! Before the dazed eyes of the men fighting to hold back the sea, the leak suddenly increased!

Hurriedly the engine room was ransacked for emergency securing gear. A bit of planking, two inches thick, once used somewhere as staging, was dragged up, and with other improvised material, jammed in against the door. The leak was at last reduced to an insignificant trickle.

For the first time then since the *Paulding*, only a few minutes before, had crashed into the *S-4*, the panting

men aft were for the time safe from immediate death.
But what faced them? Drenched, most of them, in icy
sea water; packed now, 34 men in two small compartments with hardly space to stand comfortably between
the engines; with no place except wet and oily steel
plates to lie down; with no blankets, no bunks, no heavy
clothing, no air except that fouled already from three
hours submerged operation; with no means of getting
rid of the carbon dioxide continuously exhaled from
their own breathing and poisoning the air; with the
steel shell forming their prison firmly gripped in the
freezing water of the deep sea, soon chilling the engine
room to 34° F., the trapped crew, powerless to do anything, began their weary wait for help from the world
above. But long before the first sign of that help came,
in the foul air inside the crowded stern, they had all
lapsed into unconsciousness.

On the surface, from the *Paulding*, down by the
head now and seemingly in danger of sinking herself,
the radio began to crackle:

COMMANDANT NAVY YARD BOSTON.

    RAMMED AND SANK UNKNOWN SUBMARINE OFF WOOD
END LIGHT PROVINCETOWN.   PAULDING.

From Boston to the Submarine Base in New London, the Navy Yard at Portsmouth, and the Navy Yard
at New York almost immediately went identical tele-
grams:

SUBMARINE REPORTED SUNK AT WOOD END NEAR
PROVINCETOWN BY COAST GUARD DESTROYER. SEND ANY
LIFTING APPARATUS. RUSH.

The collision occurred at 3:37 P.M. By a few minutes after 4 P.M., in New London, Portsmouth, and
New York, action had started. In New London lay
the *Falcon*, the only salvage ship the Navy had in the
Atlantic, part of her crew ashore that Saturday afternoon on liberty. Hastily the word was broadcast around
New London recalling the liberty parties. At 6:10 P.M.,
with her crew gathered from far and near around the
town, the *Falcon*, carrying Rear Admiral Brumby, flag
officer of the squadron to which the *S-4* belonged,
sailed for Provincetown, 120 miles away. From Portsmouth, at 7:30 sailed the *Bushnell*, mother ship of the
S-4 and her sisters; from Boston, several destroyers and
tugs; a few hours later, from New York, six pontoons
in tow of other tugs—the same pontoons used two years
before to lift the *S-51*; while, from Norfolk, sailed the
U.S.S. *Wright*, carrying on her deck four more pontoons, the remainder of the single lot of ten that the
Navy owned.

Meanwhile, during the night, over the road by automobile from Newport to Provincetown went the most
important thing of all—all the divers in the vicinity—
three men, Eadie, Carr, and Michels, veterans of the
S-51 salvage, with eight others of less experience.

And so, while that brief December day drew to its
close and the *S-4*, silent now, lay at the bottom of the
sea, from Maine to Virginia everything that the Navy

U.S.S. "FALCON" WITH A SALVAGE PONTOON ALONGSIDE

had or could hire in the way of men and materials for salvage and for rescue was starting for Province-town.

Boatswain Gracie of the Coast Guard, on the scene with his surfboat promptly after the collision, dropped a grapnel and commenced to sweep the bottom over the spot indicated by the bubbles of air and traces of oil escaping from the *S-4*. For four hours back and forth over that spot, first in the twilight, then in the darkness, Gracie worked with his grappling hook and, in spite of a bad sea, rode the waves with his drag astern till finally, at 8 P.M., he made a hard strike and clung to it in his boat as the hours dragged on and craft of various types began to arrive. But at 3 A.M., still await-ing the arrival of the *Falcon* and the divers, the grap-nel gave way and his boat went adrift. Undismayed, Gracie boarded the *Bushnell* (which had shortly be-fore appeared from Portsmouth), borrowed better grap-pling equipment and grimly went back with the surf-boat to his task, for there was not a small boat the Navy had in any of the ships in the flotilla gathering now around Provincetown that could live and work in that sea.

In freezing spray and in the darkness Gracie lowered his new grapnel and, under the searchlights of two Navy minesweepers, recommenced dragging. Dawn came, he was still at it. At 7 A.M. on Sunday morning, the *Falcon* steamed up from the Cape Cod Canal, took aboard the divers already in Provincetown, and stood by outside prepared to dive. But until a line was hooked into the *S-4* to guide the divers down to her, putting men over the side was out of the question. And while the fleet of nearly a dozen ships anchored near by, men and officers eager to go into action, had to stand by chafing idly, Boatswain Gracie, who since eight the previous evening and all through the long December night had clung to the wreck, swept now endlessly back and forth across it in his surfboat, trying to hook it again.

Finally at 11 A.M. his grapnel caught. Carefully, so as not to lose that precious grip, he buoyed off his line with an empty gasoline drum, while the *Falcon*, with something to work to at last, maneuvered to windward, dropped anchor, then veered cable to bring herself over the spot and took aboard the buoyed-off line. For proper work the *Falcon* required from four to six heavy mooring buoys laid out in a circle to hold her in position against wind and sea, but there was no time to lay out the buoys or to plant the anchors. Instead, two minesweepers, the *Lark* and the *Mallard*, anchored one off each quarter of the *Falcon*, and to each of them the *Falcon* ran a hawser to hold herself over the wreck as best she might.

At 1:38 P.M., twenty-two hours after the collision, Tom Eadie, chief gunner's mate, was hoisted over the *Falcon's* side as she yawed and pitched to the head seas, dropped into the water, and slid swiftly down the grappling line. Within a minute his lead-soled shoes landed with a clang high on the chariot bridge of the *S-4* between the two periscopes where the grappling hook had caught. The water was murky, the light dim, the cross current bad. As Eadie clambered down, the quiet of the deep sea broken now by the banging of his weights against the steel hull, he thought he caught coming through the water from forward a signal. As he landed on the deck, he was sure of it. Over the slewed gun, across the torn deck, Eadie went forward, following the sounds. They came from the torpedo room. At the torpedo room hatch he stooped and banged the cover. Immediately from within, strong and distinct, came six raps, repeated each time Eadie tapped. Six men alive in the torpedo room! Promptly from the bottom of the sea, the diver reported this over his telephone.

With a final rap for encouragement, Eadie went forward to check conditions at the bow, then aft over the wrecked deck to the conning tower. He rapped there. No answer. Aft again along the undamaged deck as far as the steel hatch over the engine room where he rapped again. But there was no response from anyone astern.

What to do?

On the *Falcon's* deck, listening to the diver's reports, were gathered Rear Admiral Brumby, in whose hands lay the final decision; Captain King, lately in command of the Submarine Base at New London and two years before senior officer in command of the salvage operations on the *S-51*; Commander Strother, a submarine officer of long experience, to whose division the *S-4* belonged; Commander Saunders, long engaged in submarine design and construction; and Lieutenant Hartley, captain of the *Falcon*, an expert in diving and salvage work. By Brumby the decision had to be made, guided, as his judgment dictated, by the advice of his subordinates.

Brumby knew from the *Paulding*, confirmed now by Eadie's report, that the damage was in way of the battery room; he knew that six men were alive forward in the torpedo room; and he knew that from aft, where most of the rest of the crew must have been, there was no answer to signals.

Below, built into the *S-4*, were two entirely separate emergency air lines with connections outside her conning tower, intended only for use in disaster—a salvage air line leading only to all the ballast tanks; and a compartment air line opening only into the crew compartments on the boat. To which of these two emergency connections should the next diver hook the first air line—to the crew compartments, or to the ballast tanks?

Carefully the situation was canvassed. If only one compartment was flooded, blowing ballasts with external air ought to float up the boat; if two or more compartments were flooded, that was hopeless. Still, if the men forward with the boat going down by the bow and the water therefore tending to rush into the torpedo room first, had succeeded in closing the forward door of the damaged battery room, there was no apparent reason why the rest of the crew had not been able to do the same with the after battery room door, thus confining the water to one compartment.

That no sounds came from aft more probably indi-

cated, so Brumby thought, not that the stern was
flooded, but that so many men crowded in a small space
aft were, after twenty-two hours, either unconscious or
so weak from bad air that they could not answer. If so,
prompt lifting of the stern was all that would ever save
those aft; the men forward seemed strong enough to
last till that was tried first. This decision—to blow ballasts first and try to float up the boat or at least its undamaged stern—concurred in by those present, but
which turned out to be wrong, because two compartments were flooded and in addition a ballast tank was
ruptured, was promptly put in execution.

At 3 P.M., Bill Carr, a chief boatswain's mate now,
went over the side carrying the salvage air hose for the
ballast tanks. He passed Eadie who was on his way up.
Connecting up the hose proved difficult; Carr spent
90 minutes in icy water on the bottom working through
the little hatch in the side of the superstructure before
the hose was finally coupled and some air blown
through to test the job.

Then, with Carr off the boat, the *Falcon* was hauled
a little to one side to avoid being struck by the rising
S-4, and blowing commenced. At full speed for an
hour the compressors on the *Falcon* rammed down air
through the hose into the ballast tanks of the *S-4* while
on the surface hopeful eyes watched and waited. Then
air began bubbling up, in quantity increasing till it
equaled what the compressors were sending down, and
sounded the knell of that hope. What water could be
expelled from the ballasts evidently was gone, the air
was now simply blowing out somewhere as fast as it

went down.

The *S-4* did not rise. Apparently more than one compartment was flooded. If that were so, presumably then the crew aft, for some reason or other, had not succeeded in closing any doors, and their silence meant that they had all drowned. There was no longer any hope except for the six men forward.

Meanwhile, as night drew on, conditions on the surface, never good, had changed for the worse. The long-expected northwest storm finally broke and kept on increasing, with the wind blowing a gale, and the *Falcon*, swinging from her flimsy moor, was yawing erratically from side to side and pitching heavily. It was evident now that the *S-4* could not be quickly raised by anything within the power of men; that diving, which was fast becoming impracticable in that sea, would soon for days be wholly impossible; that, to keep the men in the torpedo room alive till the storm blew over and diving could be resumed, a hose would have to be connected immediately to the compartment air line to feed air into the torpedo room.

Hastily the *Falcon* was hauled back into diving position, centered as well as possible over the *S-4*. At 8 P.M., in the darkness, with heavy spray breaking over the *Falcon's* rail as each wave hit her, and freezing immediately on the decks, Fred Michels—chief torpedo-man, third of the trio of *S-51* salvage veterans, over six feet tall, strongest and most experienced of the divers left on the *Falcon*—went over the side, carrying with him the air hose and, to illuminate his job, a powerful submarine light.

Never before had diving been tried in such a sea; probably never again will it be attempted. On the heaving *Falcon*, Lieutenant Hartley did his best to keep the manila descending line up and down over the *S-4*, but the violently yawing *Falcon*, held in position only by lines to two similar ships yawing as badly as she was herself, made it impossible. Michels, sliding down the rope to the *S-4*, hit bottom, not on the submarine but in deep mud somewhere off to one side of her, and immediately found himself buried to his waist and wholly unable to move or to extricate himself. He phoned up his difficulty. On the *Falcon* it took 13 men hauling on his lifelines to drag him clear of the mud and up into the water once more. A second attempt to land him on the submarine was more successful, but unfortunately, instead of landing near the conning tower where the descending line was attached and where he was to connect the hose, he was dropped this time in the midst of the wreckage left by the *Paulding*, where the slack of his own air hose promptly fouled. Before Michels could clear it, he felt himself irresistibly dragged to the deck by the tightening coils across his back and in a moment found himself sprawled out face down on the steel hull of the *S-4* with his own lifelines, as the *Falcon* yawed back and forth on the surface, tangling above him on the wreckage in a web from which there was no escape!

Helpless, unable to signal, unable to talk intelligently with the air roaring through his helmet, Michels managed after nearly an hour on the bottom to get a message through indicating his plight. With the weather even worse, Tom Eadie was dressed again and in the middle of a December gale went overboard to try to save Michels. Disregarding the descending line now, he slid down on Michels' own air hose as a guide. Somehow between luck and skill, those on the *Falcon* managed to land Eadie directly on the submarine the first time without fouling him. Working there in that ice water for nearly two hours with a hacksaw, Eadie cut away part of the wreckage and untangled Michels. Finally, near midnight, Michels, after three hours and twenty minutes on the bottom, went up, unconscious, perhaps dead, and literally frozen as stiff as a board.

CHIEF TORPEDOMAN FRED MICHELS (CENTER)

CHAPTER III

SUBMARINE S-4 SUNK! FORTY MEN TRAPPED!

Another submarine gone! Almost stunned, I looked at that headline shrieking up at me from the paper at my feet, then, still dazed, I picked up the paper to read the details. But the news behind the headline that Sunday morning was very scanty—the *S-4* had been rammed off Provincetown by the destroyer *Paulding*, had gone down with her entire crew inside. The *Falcon* was on the way there from New London with a few divers. That was all the papers contained, except for a picture of the *S-4* and the names of her crew. Whether the men inside of the sunken submarine were alive or dead, no one knew.

I gazed at the black type with a heavy heart. Who ever would have imagined that the tragedy of the *S-51* would be so soon repeated?

What it meant to be caught at the bottom of the sea in a smashed submarine, I knew from first hand contact with what had happened to the crew of the *S-51* when the sea poured into her. Unable to get any doors closed, her men had been overwhelmed by the inrushing waters. While diving on her, we had come across their floating corpses from one end of the boat to the other; all save two unfortunates, who, locking themselves into a small watertight trunk to escape the flood, found themselves unable to get out again and, instead of drowning quickly like the others, had slowly suffocated from lack of air before help came.

Now it had happened again to the crew of the *S-4*. Had they had any better luck in getting the watertight doors closed, before the rushing water drowned them like rats in a trap, than had their shipmates on the *S-51?* Judging from what was in the paper, nobody knew—yet.

Could I do anything about it to help, I wondered? I was a civilian, no more concerned than any other of the millions of Americans who on that Sunday morning were reading the same sad news. But, unlike most of them, I knew something of salvage and diving and might perhaps lend a hand at rescue at Provincetown if by any chance anyone there inside the *S-4* was still alive. With that in mind, I dropped the paper, went to work on the telephone, got the Navy Department in Washington by long distance, and offered my services as a civilian volunteer. They were promptly accepted, and I was directed immediately to report to the Commandant of the New York Navy Yard for dispatch to the scene of the wreck.

It didn't take long to grab out of the attic a battered uniform cap, my old Navy overcoat (the stripes and gold buttons had long since been cut off it), toss all the heavy socks I had into a bag, and get over to the well-remembered Brooklyn Navy Yard to report for transportation to Provincetown.

I promptly ran into a snag in the person of the Commandant, Rear Admiral Plunkett, late commander of

the naval railway batteries in France, those land battleships whose monster 14-inch guns in 1918 had blasted
Ludendorff out of the Hindenburg Line and made hash
of the German Western Front. There, waiting for me,
was Admiral Plunkett himself, my old commanding
officer on my last tour of duty in the Navy, looking
gruffer perhaps than ever. In my somewhat nondescript
rig, with a tarnished cap topping a naval overcoat shorn
of all insignia, I saluted him, reported for duty, and re-
quested the quickest possible conveyance to join the
rescue forces off Cape Cod.

Plunkett, tall, almost lanky, gray, bespectacled, and
walrus-whiskered, formidable in an austerity which
needed no gold lace to back it up, greeted me warmly,
told me he had already been advised by Washington of
my coming, and then, in that deep vibrant voice which
no man who ever served under him will ever forget,
threw a monkey wrench into my plans.

"I'm all ready to swear you into the Service again,
Ellsberg, and send you right up to join the *Falcon*. Just
sign these papers, then I'll administer the oath, and we'll
have you on your way in a minute."

I looked at him nonplussed.

"But, admiral," I objected, "I don't want to sign any
papers and I don't want to get sworn into anything.
I didn't come over here to join the Navy again. I told
them in Washington I was volunteering as a civilian to
help on the *S-4*, and the Department agreed. You were
here yourself, admiral, when I resigned from the Service, and you know I was all through then."

"Don't argue with me, young man!" Plunkett fixed
me with his stern blue eyes. "Do what I say. What's

Washington know about it? They don't know what's
going on off Provincetown, but I do. And you'll sign
or you don't go. You don't want to join the Navy
again, even temporarily, eh? Well, you will. Don't
think I give a damn about you, but I remember your
wife. If you go up to Provincetown in the Navy and
anything happens while you're diving, we can do something for your widow. But if you go as a civilian—"
He left the sentence unfinished, paused a moment for
his statement to sink in, then quietly concluded, "Sign
here."

Shivering a little at the implications of the "widow,"
I signed. Admiral Plunkett was right as usual, and, as
usual, giving more thought to the welfare of his subordinates than they usually gave to it themselves. In a
moment with right hand aloft before Plunkett's stern
visage, I was taking the oath, was hastily sworn back
into the Service in my old rank of lieutenant commander (except that it was in the Reserve this time).

"Now, Ellsberg, I'll get you started," announced the
admiral. "Sorry I can't give you a plane, but Captain
King got off this morning for Provincetown in the last
available plane we had. So I've wirelessed to Boston to
have a destroyer standing by there to take you across
Massachusetts Bay, and all I've got to do is to get you
to Boston. My aide's been checking train schedules;
there's a fast one leaving Grand Central in twenty
minutes. You go on that."

"In twenty minutes, sir?" I exclaimed in dismay, for
to get from the Brooklyn Navy Yard across the East
River and through five miles of Manhattan traffic to

Grand Central in twenty minutes seemed a physical impossibility.

"Don't worry; you'll make it all right," assured Plunkett grimly. "I've a Navy ambulance below waiting to take you through; if I know that driver, he won't need twenty minutes!" The admiral took my hand, gave it a friendly squeeze, and with a somber smile waved me out. "Get underway now, and good luck to you, Ellsberg, on the *S-4!*"

Again Plunkett was right; the driver landed me in Grand Central with minutes to spare. Behind clanging bell and shrieking siren, I clung to a stretcher in the back of an ambulance while an ambulance driver who must have learned his trade amongst the shell holes on the Western Front steered the course, and startled police officers far ahead held back the traffic to let him through regardless of red lights or green.

Early evening found me on the waterfront at the Boston Navy Yard, shivering in a piercing December gale in spite of my heavy overcoat. Alongside a dock there, the destroyer *Sturtevant* was waiting for me, steam up and safeties popping, ready to shove off. At the gangway, muffled to the ears, stood Captain Simmers, manager of the yard, with various assistants who had been loading emergency stores on the boat for transport to the rescue fleet. Simmers showed me an inventory list of the stores which I could see littering the destroyer's decks and asked me if I had any suggestions for additions.

I scanned his list—woolen underwear by the bale,

boots, air hoses, lead weights, diving gear, miscellaneous supplies of all kinds—everything seemed to have been thought of but one item.

"This is December, captain," I said, "and you're nearly frozen through just standing here. It'll be hell on a man diving in ice water out there in this weather. He'll need some medicine to thaw out on when he comes up. Got any whisky in the yard?"

Simmers shook his head.

"Not since Prohibition. Not even for medicine any more."

"Well, in a pinch, pure grain alcohol'll do instead," I replied. "You carry that for refilling compass bowls, I know. Send us out a couple of five gallon cans of that. We'll mix 'submarine cocktails' with it the way we used to do on the S-51, after a cold dive—half a pint hot coffee, half a pint grain alcohol—marvelous how a pint of that mixture will unfreeze a cold diver, captain."

"It ought to," agreed the manager. "We'll see that a couple of cases go out on the next boat. Anything else?"

"No, sir; that's everything." I saluted and scrambled up the narrow gangplank. With a blast of her whistle, the *Sturtevant* cast off her mooring lines and the water churned sharply into foam as she slipped out into the stream.

The weather in Boston Harbor was bad in spite of the sheltering islands on both sides, but once clear of Boston Light it promptly got much worse, when the cold wind sweeping down from the northwest struck us. In the gathering darkness, the sea was rapidly building up. As we drew away from the coast, the unobstructed sweep of the wind increased and the seas driving before the wintry gale mounted higher and higher. The *Sturtevant*, which was soon forced to slow down sharply, wallowed along, rolling as only a destroyer can, and pounding heavily into the waves. Her skipper, as best he could, nursed her along, but still a crossing she might easily have made in two hours in better weather took well over twice that, and it was past midnight when in a fierce gale the laboring *Sturtevant* finally neared Cape Cod.

Somewhere in the open sea off the tip of that cape the *Falcon*, according to late reports in Boston, was moored over the wreck, working with her divers, and to the *Falcon* I wanted to go. But the *Sturtevant's* skipper could not lay his destroyer alongside the *Falcon* even in better weather, and he had no small boat which he cared to trust overboard in that storm to take me to her. So instead of looking for the *Falcon*, the *Sturtevant* kept on considerably to the southward of her, directly for the shelter of Provincetown Harbor, in which haven I soon saw the lights of every vessel in the rescue forces except those of the *Falcon*. There were the two minesweepers, the *Lark* and the *Mallard;* the submarine mother ship, the *Bushnell;* the destroyer *Paulding* with her smashed-in bow; and some Coast Guard cutters.

Apparently diving must have ceased, which was not surprising in view of the state of the sea, but that the sea was so bad that all the ships had been forced to run for shelter, rather than take a chance on dragging anchors *en masse* and possibly fouling each other, had not before occurred to me. However, there they were, save for the *Falcon*. Since the *Falcon*, with her divers and her diving gear, my destination, was still somewhere outside in that howling waste of waters, I had to get a boat and get out to her.

It was after 1 A.M. when, the *Sturtevant's* anchor having rattled out, I watched anxiously from the bridge alongside her skipper as the destroyer's blinker lights started to twinkle through the night, calling the *Bushnell*, by far the largest vessel present, then the *Lark*, the *Mallard*, and finally the Coast Guard cutters, asking for a boat to take me out to the *Falcon*. But from each vessel in succession came the same reply, a negative. Not one ship had a boat she would risk sending outside the harbor.

"No luck, Ellsberg," reported the skipper sympathetically. "And I can't say I blame any of 'em. No Navy boat I ever saw could make her way that far to windward against this gale; she'd swamp sure. Better spend the rest of the night here with me and if the gale's gone down any in the morning, maybe you can get a boat then."

I shook my head. I had come a long way to join the *Falcon*. What the situation was inside the *S-4* I didn't know; what the diving situation was on the *Falcon*, I didn't know either; but if the men in the *S-4* could in any way be aided by the presence of another diver, I was not going idly to stand by only a few miles off if I could by hook or crook get aboard.

"Put me aboard that nearest Coast Guard cutter,

skipper," I asked. "Maybe they can fix me up."

In the black night, muffled again in my overcoat, at
nearly 2 A.M. I was transferred from the *Sturtevant* to
the cutter, and, mounting her side ladder, soon found
myself in the captain's cabin, to be greeted not only by
her skipper but also by the Coast Guard Supervisor for
the whole New England district. Briefly I repeated my
request for a boat.

"It can't be done," answered the supervisor. "We
haven't a boat on the ship I'd risk outside in this storm."

"Yes, captain," I agreed, "but I've got to get out to
the *Falcon*, and you people have got to help me. I know
as well as you that regular ships' boats are out of the
question in this gale and that no pulling boat's crew
could fight their way so far against the sea, but your
Coast Guard outfit is in the lifesaving business, isn't
it? And you've got a special motor surfboat in that
station of yours ashore that's built to go out in hurricanes. I used one of your small motor surfboats for bad
weather salvage work before, and she was a grand boat
for a storm. How about drumming up something like
that for me now?"

Here was a new idea. The Coast Guard officers
looked at each other, then nodded slowly in agreement.
The supervisor finally spoke.

"That's right. One of our motor surfboats could
make it. Where's Bosun Gracie and that boat of his
from the Wood End Station, captain?"

"Gracie and his men are all knocked out, sir," replied the commander of the cutter. "They were out
over twenty-four hours dragging for that sub, and

when they finally hooked her the second time and came in frozen stiffer'n planks, I sent 'em all ashore to recuperate. But the boat itself is still tied up at the boom, in charge of some men from around Truro further down the cape that've been moved up here as reliefs."

"Fine!" exclaimed the supervisor. "Get that relief bosun in charge of her up here immediately."

While we were waiting, in came the cabin messboy with some hot coffee for which, being somewhat chilled, I was duly grateful. The coffee and the allusion to the frigid state in which Bosun Gracie had finished his labors brought to mind the conversation I had had early that evening with Captain Simmers.

"For a real case of exposure, a man needs something stronger than coffee to warm up on, captain," I murmured as I sipped. "I tried to get some whisky as I came through the Boston Yard, for the divers out here, but whisky's not on their books any more, and the best they could do for me is some grain alcohol they're sending out today."

"Whisky? You say you want some whisky for your divers?" The senior Coast Guard officer, who had been sitting silently by while I drank my coffee, apparently somberly considering the plight of the S-4, suddenly manifested an unusual interest. "Whisky, eh? Well, you've come to the right place for that. There's something at least I can help you fellows on and no argument. My district base is loaded down with Scotch we've taken from captured rum-runners, and I can honestly assure you it's the genuine stuff, right off the boat, which is more than any bootlegger nowadays really

knows about what he's selling. Don't drink that grain
alcohol Boston's sending you. I'll get you Scotch instead. That'll be a little something the Coast Guard
can do to help the Navy out here. How much do you
want?"

"Thanks, captain. A couple of cases to start on'll be
plenty. I'll let you know later about more." I smiled a
little grimly at the situation. Queer things happened
under Prohibition. The *Paulding*, a Coast Guard rum-
chaser, had sunk the *S-4* (though apparently through
no fault of hers); now the Coast Guard was going to
supply some captured rum for the divers working on
the ship they sank.

A knock at the cabin door, and in came the relief
bosun, a small man lost in an oversized peacoat he had
grabbed somewhere to shield him from the cold, and
fidgeting nervously over the knit cap in his hands as
if uneasy in the presence of more rank than he usually
saw in a dozen years.

"You want me, sir?" he asked, looking inquiringly
from one to the other of the two Coast Guard captains.

"Yes," answered the skipper. "Here's a naval officer
wants to go out to the *Falcon* right away. Can you put
him aboard in that surfboat you took over from
Gracie?"

"Aboard the *Falcon?* Now? No, sir!" replied the
bosun decisively. "Ain't a chance. I can get out there
all right, if I can keep the engine going, because that
lifeboat's unsinkable, and I reckon I'll keep her right
side up, but I can't land nobody on the *Falcon*, captain. The sea running out where she lies 'll smash the

boat to matchsticks if I lay her alongside!"

That sounded like the end, but I caught at even the slight straw of encouragement in the bosun's answer.

"That suits me, captain," I injected hastily. "Get me out there and I'll take my chances on getting aboard. So long as your bosun thinks he can get out there, let's go. Maybe he can put me aboard. But if he finds he can't land me, I'll come back."

My two hosts looked a little dubious at this proposal, the more so as, on further questioning, the bosun stuck by his opinion that trying to go alongside meant sinking the boat, but I was insistent that a trial be made. Ordinarily I would never have got the boat, but the district supervisor, feeling that, since the Coast Guard was involved in the *S-4* tragedy, they must lean over backward to help the Navy, finally, against his own better judgment and with great reluctance, ordered the trip.

At 2 A.M. on the morning of Monday, December 19, in a Coast Guard lifeboat, I shoved off from Provincetown for the last leg of my voyage and was shortly outside of the harbor. If the going had seemed bad an hour before while aboard the *Sturtevant* running with the sea on her quarter, now when, clear of the harbor, that small boat met the full fury of the gale head on, it seemed impossible to live. The boat tossed wildly in the waves, rising with sickening speed as the crests lifted the bow high, then lurching dizzily down at a sharp angle into the troughs as the waves went by. Solid water came over the stem in sheets which would promptly have swamped an ordinary boat, but the decked-over forecastle and self-bailing cockpit of this lifeboat kept her buoyant and afloat while one after another in endless succession she drove through the darkness into the long storm waves. Slowly against wind and sea we ran to the westward for the wildest ride of my naval experience. Crouched in the stern-sheets alongside the bosun, I quickly was coated with a sheet of ice from the waves splashing aboard and in that bitter weather promptly freezing, while the salt spray, turning to tiny icicles in the intense cold, drove like a stinging sandblast into my face.

Constantly heaving on the tiller, alongside me the bosun steered, more by feel than by sight holding his boat head on to the seas to keep her from broaching to and rolling over, while very slowly the lights of the *Falcon* in the distance drew nearer. Possibly his exertions at the tiller kept the helmsman limbered up, but in that cramped space, motionless, with nothing to do, I felt myself gradually becoming hardly more than a block of ice freezing to the gunwale.

The bosun, noticing my plight, motioned me to get down with the engineer in the tiny shelter amidships housing the motor. With ice cracking off me in sheets as I cautiously went forward clinging to the rail to avoid being heaved overboard, I wedged myself inside the engine housing, for a moment grateful for the warmth rising from the laboring engine. But I soon found I had hardly improved my position. Crouched over the engine I was warmer and protected from the piercing wind and the stinging spray, but the atmosphere in that compartment, closed to protect the engine from the flying spray, was a stifling mixture of oil fumes rising from a smoking cylinder block and of vapors from the gasoline spilling copiously from the carburetor as the wildly gyrating boat did everything but turn over.

How the engineer crouched opposite me could stand it even for a minute was incomprehensible. With the alternative of being asphyxiated over that engine or of freezing in the open cockpit, I decided to freeze, and, leaving the engine compartment, crawled slowly aft again to the sternsheets to face the storm.

Darkness, bitter cold, tumbling seas, and a roaring wind—for nearly an hour we battled them in that lifeboat to cover the three miles from Provincetown to where in the unsheltered ocean the *Falcon* lay. At the end of that hour, numbed with cold and mercilessly seared by the wind, we came at last abreast her, streaming at the end of her anchor cable paid out to full scope, heaving erratically to the long seas sweeping by, yawing wildly about that cable to her bow.

It was 3 A.M. Slowly the tossing surfboat worked its way past the port quarter of the *Falcon*, keeping a respectable distance off. The storm waves were beating fiercely against the ship, slapping her side on each roll, pouring in wide sheets from her bow each time she lifted. It was obvious that there was no chance of boarding on that side. Continuing, we cleared the *Falcon*'s stem, kept on well ahead to make sure we were clear of that madly surging anchor chain which in the darkness we could not see, and then, turning carefully, ran with wind and sea astern down the starboard side of the ship. Anxiously I scanned the weather on that side to see if it offered any lee, but there was no choice —starboard or port were quite the same and both sides were impossible. Tugging violently at her anchor, the *Falcon* was making far worse weather of it than if she had been underway and free to lift. Wave after wave roared up to smash against her bow, cascade in a terrifying mass of broken water down her sides, and disappear in a maelstrom of foam under her stern as her flat counter thudded down into the trough with a crash like rolling thunder.

We cleared her starboard quarter with our lifeboat headed down-wind toward Provincetown.

"No use!" roared the bosun, shouting to make himself heard above the screaming of the wind. "It's like I told you. Can't make it. I'm going back!"

I shook my head.

"Not yet!" I yelled. "Circle her again! We may get a lull in the gale!"

The bosun shook his head dubiously, but nevertheless he put his helm aport and we rounded the *Falcon's* stern for another run up her port side, going as slowly as we dared.

There was no improvement. Alongside that vessel, at one moment the racing crests were washing over the gunwale, the next moment the troughs were exposing almost her entire hull down to her bilge keel. No small boat, churning in the seas there, had even a slight chance of surviving an impact with the heavy sides of the rolling *Falcon* crashing down on her.

I looked anxiously at the ship as we went by. Except for her anchor lights, she was almost dark, and not a single man could I see on her decks anywhere, either to answer a hail or to give any assistance. And from the looks of the *Falcon* heaving madly in that wintry sea, anyone trying to clamber aboard her would need plenty of assistance, for her sides and gunwales were a solid mass of glistening ice from bow to stern! It was obvious enough why they weren't diving and why all the *Falcon's* consorts had gone into port, but what had happened on the *Falcon* that there wasn't even a watch visible on deck?

Once more we cleared the bow, kept on to make sure we cleared her cable, and then, rocking wildly as we came broadside to the seas in turning, managed to get about and head aft before the gale down the starboard side again. There was no sign of any lull in the storm, no vestige of a sheltering lee anywhere. Getting alongside was completely hopeless; it was useless to argue otherwise, the bosun was right.

"Head back now, sir?" he shouted.

"No!" I sang out. "Round to under her stern and head up her port side again!" I paused a moment for breath, then continued, "Pass her as close aboard as you dare without staving in your boat, and I'll jump for it on the fly!"

Clinging to his tiller, the bosun roared back:

"Don't try that, sir! You'll drown!"

"Don't worry, I'll make it! Port your helm!"

Obediently he rounded to. The boat was his responsibility, but I was not.

For the third time, the surfboat headed into the teeth of the storm up the port side of the yawing *Falcon.* With engine slowed and with only speed enough on to keep steerage way and avoid broaching, the lifeboat edged up past the stern, past the quarter, and then sheered gradually in till we were not over eight feet away from the *Falcon's* side, while poised on the life-boat's gunwale, clutching the engine house, I clung pre-cariously, watching the lifeboat crazily rise and fall in the rushing waves, one moment deep in a trough with the *Falcon's* side towering like a cliff over me, the next riding high on a crest looking down on the *Falcon's* ice-coated rail. There was no earthly chance of ever getting a grip on that rail—it was either clear it or land in the sea.

We dropped dizzily into another trough as the *Fal-con* started rolling toward us. I braced myself, looked aft at the bosun, nodded. Along came the inevitable crest, the bow of the boat heaved up, the bosun sheered in a little. For an instant only, poised on that wave, I could see the *Falcon's* glistening side below me in the darkness and the black water gap between ice-coated ship and ice-coated lifeboat narrowing. Then head first, with arms outstretched, I leaped in a wild dive from boat to ship, shot over the gunwale, and felt myself sliding spread-eagled on my stomach down the *Falcon's* icy deck!

While I was still sliding, my bag, heaved across by the engineer, came hurtling over the rail to land with a crash astern of me. By the time I had regained my feet, the lifeboat, with her engine at full speed once more, was already several boat lengths away and headed about, disappearing in the black night toward Province-town.

"Thanks!" I shouted across the water, but in the storm I doubt that I was heard. The bosun, back to me now, was only an icy blotch in the darkness against the icy whiteness of his boat, which, driving now dead before the gale, quickly vanished in the roaring night.

And I at 3 A.M., completely alone, stood at last on the deck of the *Falcon*. My journey, weirdly commenced in an ambulance behind a siren shrieking through the streets of New York, and just as weirdly ended by a sickening leap after a wild ride in a storm-tossed lifeboat through the gale-swept sea off Cape Cod, was over. What now of the trapped seamen inside the *S-4?*

CHAPTER IV

A LITTLE shaken by my leap, I clung an instant to a near-by superstructure stanchion to steady myself, then started forward along the deserted deck. Near the bridge, a quartermaster hastily descending from the chart house met me, apparently recognized me from my last cruise on the *Falcon*.

"Where's the skipper?" I asked briefly.

"Turned in, sir, dead to the world. So's everybody else on deck. They're all knocked out. We've had a wild night diving. You remember big Mike? Well, he's in the recompression chamber now. Whether he's dead or alive, I don't know. Bill Carr and Tom Eadie have been working on him in the tank since he came up frozen stiff a couple of hours ago, and they put the pressure on him trying to bring him to. It looks bad for Mike."

"How about the sub?" I asked. "Anybody alive in her?"

"Yeh. Six of 'em, locked in her torpedo room. Mike went down in the storm to take 'em an air hose, but the gale shot everything to hell, tangled him up in the wreckage below, an' he never made it. God help 'em now!" The quartermaster knew no more.

I scrambled across the deck to the starboard side where, built into the ship, lay the recompression tank, a huge steel cylinder for treating divers under air pressure. I swung open the outer door, a round steel casting like the entrance to a bank vault, only smaller, crawled through the opening, swung the heavy door to again behind me, and then, opening an air valve inside, let compressed air pour into the lock in which I stood. It was warm, almost hot, in that chamber, so while waiting for the pressure to build up, I tossed aside as useless encumbrances now my ice-coated overcoat, hat, and mittens, and stood before the inner steel door, pushing it occasionally to see when the pressure on my side had built up enough to allow me to swing back that door.

Finally the air built up sufficiently to equalize, the door swung back, and I shouldered through into the inner chamber. My heart sank at what I saw.

On the deck, naked, stiff, unconscious, lay Fred Michels whom last I had parted from long months before at the Navy Yard in New York. Working over him, one on each side, chafing his muscles, rubbing him with hot towels, were Tom Eadie and Bill Carr, striving desperately to bring him to. For over two hours now they had been at it. Without a word, I dropped to my knees alongside to help.

Neither Carr nor Eadie showed any surprise at my sudden apparition at that unearthly hour. Nothing in connection with the *S-4* would ever surprise them again. Wearily they moved over a little, gratefully accepted my help on Michels. But while I worked on the still form before me—kneading muscles, chafing,

EXTERIOR OF A RECOMPRESSION TANK

rubbing, trying to work a little warmth into that stiff shape which for over three hours had lain motionless, trapped in a web of air hose on the *S-4's* broken deck at the bottom of the sea, literally submerged in ice water greedily sucking every vestige of warmth from his body—I gathered bit by bit from Carr and from Eadie what little they knew of the *S-4*, how matters stood with her then.

Eadie had learned on his first dive that Lieutenant Fitch and five torpedomen were alive in the torpedo room forward, but no signals had come from the thirty-four men aft. Then Carr, making the second dive, had coupled up a blowing hose, but the sub had failed to rise when they blew ballast tanks through the hose, and then the storm had burst. In a desperate effort to get an air hose to the men in the torpedo room before diving became impossible, Michels had gone overboard in the darkness and the gale from the madly heaving *Falcon* only to be trapped below in the wreckage. That Eadie, diving a second time, had ever succeeded in freeing him seemed a miracle to me, and now there lay Michels and diving was over and one hundred feet below us on the bottom of that storm-swept sea lay the *S-4* with six living men beyond any help—unless the sea abated.

There the story ended. Silently we three worked over Michels. Unconscious, frozen through, saturated in every vein and tissue with nitrogen from over three hours on the bottom under heavy air pressure, then as a final blow to be brought up without the decompression necessary to save him from that dread disease of deep diving, "the bends," his chances looked slim. If we ever got him thawed out, the probabilities were that pneumonia would get him if "the bends" didn't.

Chafe and rub, rub and chafe—we kept endlessly at it. Then at 3:30 A.M., Michels, whose last conscious moment had been passed at the bottom of the cold sea with his helmet pressed tightly down against the S-4's broken hull, came suddenly to, blinked his eyes questioningly, unable to believe at first whom he saw bending over him, and then mumbled in surprise:

"Why, hello, Mr. Ellsberg!"

The gale blew on. Monday morning came, no chance of diving.

Soon after the first dive, communication had been established with the men in the S-4 by Morse code, signals sent on an oscillator from the Falcon or a sister submarine, the S-8, and answered by hammer taps from the torpedo room, one rap a dot, two raps a dash. With the exception of the first taps on the hull by Eadie, all succeeding communication was from ships on the surface.

Just before Michels went down, the Falcon signaled: "Is there any gas?" To which, in dots and dashes, picked up by the Falcon's receivers, came the answer:

"No, but the air is bad. How long will you be now?"

"How many are there?"

"There are six; please hurry."

To this, the Falcon, about to lower Michels over the side in the teeth of the gale, replied:

"Compartment salvage line now being hooked up."
But that air connection was never made.

For the next two days, the gale blew on and no diving was possible. To save Michels's life, in the face of
weather which made a transfer at sea impossible, early
Monday the *Falcon*, with Michels still in the recompression tank, ran for Boston to put him in the hospital,
returning late Monday afternoon. For this temporary
departure a storm of criticism burst on Admiral
Brumby's head. And during the rest of those two days
on Monday and Tuesday—while the gale blew and the
Falcon helplessly pitched and tossed over the grave of
the S-4, while those aboard her, listening to the weak-
ening signals from the inside of the submarine clenched
their teeth and prayed for weather in which, with even
a slight chance of not killing the diver, a man could go
overboard—the storm of criticism from all over the
country rose higher and higher. It was fanned unfortunately by the misinterpretations of reporters sent to
Provincetown, who, unable to find out from the commander of the salvage fleet what was going on and
why, took his silence as something sinister, as a measure
intended to cover up the inefficiency and perhaps the
cowardice of the men on the *Falcon*.

The wildest suggestions for rescue poured in—from
the fishermen at Provincetown who knew nothing
about diving, from men and women all over the country who knew little of the sea, less even about submarines. But, with conditions what they were those

three days off Provincetown, the divers could not work, and without divers nothing was possible. Pontoons, floating cranes, trick apparatus of many sorts— all were there before the end, but in the face of that storm they might just as well have stayed in New York, Boston, and Norfolk.

Meanwhile the messages: On Monday afternoon, from the S-4 already forty-eight hours on the bottom, a question:

"How is the weather?"

"Choppy."

And later that evening:

"Is there any hope?"

Vividly do I recall, when that message came in, looking hopelessly out with Captain King at the gale raging about us, at the weather reports that it would continue to blow all next day, then King's set face as he dictated the answer:

"There is hope. Everything possible is being done."

The hours dragged on. Late Monday from the Navy Department we received by radio the following order:

"Transmit if possible the following message for Lieutenant Fitch inside the S-4:

"YOUR WIFE AND MOTHER CONSTANTLY PRAYING FOR YOU."

A little after midnight Monday that message in dots and dashes from our oscillator went down through the cold sea, to beat against the silent hull of the S-4. No answer. All through that stormy December night at

short intervals that message in the shrill note of the oscillator rang through the ship, vibrated through the
water:

"LIEUTENANT FITCH: YOUR WIFE AND MOTHER CON-
STANTLY PRAYING FOR YOU."

And on the *Falcon*, broken-hearted men, who best
knew the plight of Fitch and his companions, prayed
for a change in the weather.

Six-fifteen A.M. Tuesday morning. Sixty-three hours
on the bottom. From the *S-4* at last an answer in code:
"I understand."

The storm blew all day Tuesday.

Wednesday at last the sea moderated and the *Falcon*
moored for diving. The sounds from the *S-4* had long
since ceased. And then another blow. Both the manila
descending line the divers had used and the ballast tank
air hose, buoyed off at the surface, had chafed through
during the long storm. When picked up in the waves
on Wednesday, the last remaining threads parted, and
there was no line to the *S-4*. Moored as exactly over
the spot as bearings could place her, divers went overboard and floundered impotently in deep mud on the
bottom searching for the *S-4*. Though unquestionably
within 50 feet of the submarine, they never saw her.

Precious hours were lost till Boatswain Hawes of the
Falcon hooked her again in the afternoon with a grapnel. Hurriedly then divers Wilson and Eiben connected an air hose directly to the torpedo room through
a fitting improvised to go over a listening tube on the

submarine's bow, and we started alternately to pump in and vent out the air below. The first sample tested of the air coming up from the torpedo room showed it was hopeless; the carbon dioxide percentage was so high, 7%, men could not live long in such an atmosphere.

In spite of that, the ventilation was continued for hours till the air inside the submarine was purified, but inside the torpedo room no one ever revived. The tragedy of the *S-4* was over.

CHAPTER V

WHAT injuries had the *S-4* suffered in that collision with the onrushing *Paulding?* Had the destroyer so knifed into her, was her back so severed by the impact that she would break in two if we tried to lift her? The *Paulding*, lying there in Provincetown with her stem smashed and crumpled and her bow so deep in the water that she seemed in danger of sinking, was evidence enough of the terrific violence of the collision. What were the effects on the *S-4?*

To determine that, while the ventilation of the forward compartment went on, I was dressed for my first dive. Cased from ankles to neck in three suits of heavy blue north woods woolen underwear and wearing three pairs of thick woolen socks, I went aft to the dressing bench looking decidedly rotund.

I seated myself on the bench on the *Falcon's* fantail —for all my clothing, chilled in the cold wind—and the dressers went expertly to work on me. First came the diving dress, a canvas-covered rubber suit. Into this stiff garment I slid thankfully, for at least its impervious texture made it an excellent windbreaker, shielding me from that piercing wind. While the dressers lifted me and the suit by its rubber collar, unceremoniously shaking me down inside, I slipped my padded toes into the feet and my bulky gloved hands down into the watertight gloves forming the ends of the diving sleeves. With a final wiggle of fingers and toes to slide everything home, I settled back again on the bench, and the rest of my ponderous regalia was hastily put on. A pair of 30-pound lead shoes were strapped over my feet, an 80-pound lead belt draped round my waist, a copper breast-plate tightly bolted to the rubber collar of my dress to make a waterproof seal, a telephone headset buckled on over my ears, a massive diving knife hooked to my belt, and I was ready to test out. Hurriedly my telephone was checked, my air hose tried. Both worked satisfactorily, and down over my head came the helmet. A quick twist on it from the tender to lock the screw joint, and I was ready. Completely shut inside my suit, I opened the valve on my air hose a trifle to get some air to breathe, and immediately my rig ballooned out under the slight pressure.

Two husky dressers seized me, one by each shoulder, and helped me rise. Staggering under a load of 200 pounds of lead and copper, I walked unsteadily, supported by the dressers, shuffling my heavy shoes across the deck to the *Falcon's* rail. With an effort, I lifted my bulky shoes a few inches onto a steel stage dangling from a boom overhead and gripped the steel bails of the stage to steady myself. Then the tenders let go.

A bosun's pipe shrilled, a winch rattled into action, up went the stage, out swung the boom, and the next instant I was swaying erratically, with the stage outboard of the bulwark and only the sea beneath me. Again, muffled by the noise of the air whistling through my helmet, I heard the bosun piping, and down into the sea splashed the stage. Instantly my canvas suit, which had in the air been grotesquely swelled out, collapsed like a punctured balloon, the folds of canvas-covered rubber now pressing in tightly against my frame, leaving only a little air space inside over my chest where the breastplate held it out.

Down a few feet, and the stage stopped. Peering out of my faceplate through the water, I could see alongside me, the rounded red hull of the *Falcon* sloping away to her keel, with here and there a few barnacles and some moss. Over me was the surface, undulating like a silver sheet, sharply dividing the world I was now in from the normal world of men—a billowing sheet, crested, wrinkled, flecked with driving foam, letting through the dull December light but completely shutting off any view above as effectively as a ground glass screen.

I was alone now and truly in another world. Gone was the sense of being crushed down by all the weights I carried. Completely immersed in the sea, I was now buoyed by it, and the lift on my helmet as it tended to float up in the water took from my shoulders the burden of my bulky belt. Quickly I adjusted my air valve and the exhaust valve on my helmet to give me some negative buoyancy, enough to prevent my floating upward off the stage, and then through the telephone transmitter I sang out:

"Topside, there! All ready! Lower away!"

I felt the lifeline and air hose tied to my suit at the breastplate tauten up as my tender took in the slack, then a far-away voice sounded through the transmitters over my ears:

"O.K. Step off the stage!"

I stepped off, the stage was promptly hoisted, and there I hung, dangling on my lifelines, with nothing below me now but the bottom of the sea, and in front of me only a manila line tied somewhere below to the *S-4*. I grabbed the line and wound my canvas-covered legs round it. The tenders started to lower.

The light, never very good, quickly faded as I sank. A few fathoms down, the *Falcon* disappeared from view, the surface faded away, and around me was nothing but water into which, a little above me, the manila line down which I slid seemed to dissolve, and out of which, beneath my feet, more line seemed to materialize constantly out of nothing.

Down I slid endlessly, through a world composed of nothing but dimly lighted water and a few fathoms of manila line. The pressure increased, breathing became harder, my eardrums dilated painfully, I swallowed hard continuously to relieve them. The cold water started to strike through to my skin in spite of my many layers of protective wool, to add an intense chill to my other discomforts.

And still down I went through the seemingly bottomless depths, through an unearthly quiet broken only by the air whistling through my helmet and then out, leaving behind me a trail of air bubbles streaming from my exhaust valve, rising, expanding, breaking up in the water as they rose like huge clusters of grapes spiraling magically upward on some gigantic vine.

But I had no time for watching bubbles. My eyes were down now, always down, peering through my

faceplate at the manila descending line which was my
guide. The line started to slope away more sharply toward the horizontal; I took a firmer grip with my legs
to hang on to it. And then vaguely forming out of the
water below me was the S-4, an imponderable shape at
first, which, like the rope down which I slid, seemed to
be materializing fantastically out of nothingness. A few
more fathoms down the line, and I stopped, shouted
into my telephone:
  "On the bottom!"
  There I was at last, on the S-4. There she lay before
me, silent, motionless, huge in mass as I stared at her,
her bulk strangely magnified by the water. I was standing on her very bow, in the thin triangle where sides
and deck met stem, a precarious perch with no railing
for support. A little dizzy from the pressure, I paused
a moment, clinging tightly to my guide line, while I
readjusted air valves to suit conditions on the bottom.
Then, checking my lifelines to make sure they were not
fouled round the descending line but were floating
clear, I let go the manila rope which had been my guide
till now. Signaling on my lifelines for more slack, I
cautiously walked aft. The deck was level, the submarine had neither heel nor trim. The visibility was fair;
I could see perhaps ten feet; beyond that, like an opaque
screen, the water shrouded everything.
  Aft I went, pushing slowly through the water in the
queer walk of the diver, resembling nothing so much
as a slow motion picture gait, the clang of my lead-
soled boots ringing out metallically against the steel
deck. Beneath me was the torpedo room—there at my

feet was the round hatch on which Eadie had first hammered, on the under side of which Lieutenant Fitch
had banged out his answering signal. All was quiet inside the S-4 now. No man, with who can say what
desperate hope surging through his breast, sprang up
beneath me inside that torpedo room to beat a frantic
tattoo in answer to the clattering of diving boots against
that sunken hull. It was too late now for anything,
except, if possible, to lift that steel coffin with Fitch and
his shipmates to the surface again for burial. But could
we lift it, or would it break in two?

I continued aft, lifelines trailing behind me through
the water, helmet bent forward, eyes fixed to the deck
looking for damage. So far, nothing. The S-4, silent,
immobile, bow diving fins still trained out, rested there
on the bottom as erect, as unhurt as if she were simply
bottoming for practice and might at any minute start
up her motors and swim gently upward.

The deck widened still further. I was perhaps 70 feet
aft and increasing my pace, when before me the deck
suddenly vanished, torn completely away! In a ragged
tangle of torn steel, the deck ended where the *Paulding*,
evidently riding across, had ripped the superstructure
and the deck clean off the submarine. Peering over the
edge of the broken deck, I could see below me the cylindrical hull. Cautiously so as not to cut open my diving suit on any of those jagged plates, I clambered down
to the hull below, then continued aft again, eyes glued
now on that round steel shell, searching for the rupture
which had flooded the inside of the submarine. But
there was no opening, no gash in it anywhere, just that

smooth round hull beneath my feet with the superstructure wiped clean off, except to port where in an ugly-looking tangle lay the twisted remnants of the superstructure in which Michels had been trapped.

For 20 feet more I kept on, scanning the exposed hull. Then there rose abruptly before me, torn and jagged again, the continuation of the deck, and just abaft that, looming massively through the water, the S-4's forecastle gun, slewed drunkenly to port, breech high above the starboard rail, muzzle down, as the careening side of the *Paulding* hurtling by had twisted it from its normal fore and aft position and jammed it down on the deck.

Carefully keeping to port, I crawled up over the broken deck, dragged my cumbersome rig up on the depressed gun muzzle, holding my precious air hose high to avoid fouling it in that mass of torn steel, and slid down off the gun to land on the undamaged deck.

Inwardly thankful for the safer footing, as fast as the drag on my lifelines would permit breasting through the water, I went on aft till, standing sharply like a precipitous island, the conning tower of the S-4 rose in my path, so far as I could see wholly undamaged and showing no definite sign of contact with the *Paulding*. Certainly there could be no damage to the boat from there aft.

Puzzled, I stopped. Where was the *hole* in the submarine which had sunk her? The damage to the deck and to the superstructure which I had seen was in no way vital. There must be a hole torn in that sub somewhere. That hole I was supposed to find, to examine carefully before the salvage operation proceeded. Certainly it must be forward of me; somehow I had missed seeing it. Signaling on my lifelines to start taking in slack, I turned, retraced my steps to look for it again.

Once more I clambered over the slewed gun, dropped down over the broken deck edge onto the cylindrical hull, and started slowly forward, staring sharply down through my faceplate at the smooth round plates below me, searching from starboard to port for that hole.

A few steps forward, and then, as if a fog had suddenly rolled in, the submarine disappeared—from my waist down I stood in a cloud of mud; in no direction was any part of the S-4 visible!

I stopped instantly, all sense of direction lost, fearful that, if I took a single step the wrong way, I should go sliding overboard from the rounding cylinder on which I stood. Where was the fore and aft line now along the hull of the sub that I must follow if I were to stay aboard her? Perplexed, I looked around. The gun and the conning tower astern of me had faded away, dissolved so to speak in the translucent water. They could not serve as guides. I reflected. The sub, I knew now, lay on the bottom on a north and south line. If only I could get a compass direction, that would fix it. It was late afternoon, the sun was in the west. I looked up hopefully through the water, trusting to discover which way the light of the sun was coming. Useless. Down in the depths there was only a uniform semi-twilight pervading everything, coming from nowhere. Another idea. Before me, starting in a long curve from my breastplate, I could see a few fathoms of my lifeline and air hose rising toward the surface and the *Falcon*. I might get my direction from that. Which way was the *Falcon*, east or west, north or south, from where I stood? I stared through my faceplate at those lines, undulating gently till, not so far above me, they also dissolved in the water. There was no answer. The *Falcon* might be in any direction.

I was completely lost. I looked down at the clouds of mud billowing there in the water like clouds in the sky, blotting out my legs, blanketing the submarine. What had happened? Apparently in my passage aft along the sub I must have stirred up a fine layer of mud with which the sea had coated the *S-4*. Now going forward again, I was caught in the resulting fog. Somewhere to port of me was the wreckage in which Michels had been trapped, to starboard the curving side of the hull. A step in any direction but the right one would land me in trouble, but which way in that mud-shrouded water was right? I stood there motionless, cursing myself for a fool. Why, when I was safely out of the Navy and through with diving forever, had I been idiot enough voluntarily to come back and get into—

And then it came. On the surface, the *Falcon* took a wide yaw among the waves, my lifelines suddenly tautened, jerked my breastplate, threw me off balance, and I felt myself going over sideways, in another instant to be sprawled face down on the submarine, sliding helplessly through the water over the curving hull of the *S-4*! Faster and faster I went, lead weights clattering, copper helmet banging against the submarine's shell, while involuntarily I tried to dig my gloved fingers into the steel plates beneath me, to get a grip on something, anything, to stop my fall. Useless. The plates were smooth and slippery, there were no projections. With increasing speed, I shot overboard and started to drop vertically, still clawing wildly. No use. Down I went.

Then suddenly in the water a projection flashed before my faceplate! Out shot my right hand, grasped it. I stopped with a jolt that nearly jerked my arm from its socket, to find that there before my face was what I had made that dive to find. Through the faceplate of my copper helmet I was staring straight into the hole punched through the *S-4's* side into her battery room! And the projection I was clinging to was part of the *Paulding's* steel stem, still jammed like a broken lance into the *S-4's* death wound!

Dangling there in the water by one arm, completely forgetful of my own plight, I swiftly examined that hole where it pierced the inner hull. It was a surprisingly small gash—hardly a foot across—to have sunk that ship and killed her crew, but still, aided by the increasing pressure of the sea as the sub went down, it had been more than enough to pour into the battery room a torrent which quickly spelled complete disaster.

A second look into that hole finished my examination. The damage was insignificant as affecting the strength of the ship to stand a lift—no danger of that trifling hole causing it to break in two when our lifting gear took a strain.

With that, I swiftly forgot all about the submarine's troubles and came back to my own. We could lift the submarine all right, but how about myself? I had broken my fall; what could I now do to lift myself up on deck? I tilted back my head and looked out through the top port of my helmet. Then cold fear suddenly gripped me. A stream of bubbles was pouring upward through the water from my hand! The jagged steel to which I clung had cut open my watertight glove, and from the highest point in my suit I was rapidly losing all my air!

Frantically my lead-soled shoes beat the sides of the *S-4*, trying to get a foothold on something to support me, to allow me to drop that arm below my helmet and save my air. It was useless. On the sheer side of the submarine there wasn't the slightest toehold. I felt the sea pressing in on my chest as the air went out and breathing became more difficult. Despairingly as I dangled there I glanced up again at my cut glove, at the air bubbling away. I couldn't climb up and I couldn't hang there much longer or the sea pressure would finish me. There was nothing to do but to let go while still I had a little air left in my suit, and take my chances on the sea floor. I let go.

Down I went again through the water, faster than ever now, with little air to buoy me up, and that 200 pounds of lead and copper dragging me into the depths. I got one last glimpse of the side of the *S-4* shooting by my faceplate and then—the light went out!

I had hit bottom, but instead of stopping, I shot completely through it to find myself buried in soft mud, engulfed in total darkness, and still sinking helplessly, dragged down by my weights!

I came to rest at last, sprawled out sideways in utter blackness, to feel mud pressing in on me from all directions while the water which had leaked into my suit from my cut glove now all poured into my helmet and half strangled me. Convulsively I tried to straighten myself, to get my feet down and my head up, but with each desperate flailing of my arms and legs I could feel myself only sinking deeper through that clinging mud, lead shoes on my feet, lead belt on my waist, copper helmet on my head all equally dragging me down again with every spasmodic struggle. Was there no bottom to that mud?

Then at last I struck something hard, quit sinking, came to rest still sprawled out on my right side. With a gasp of relief, I thrust my arm down hard, intent on getting my head up, getting that water out of my helmet, only to feel my arm go full length down again into unresisting mud. There was nothing solid under my shoulders; just my body was resting on anything.

Instinctively I began to wiggle myself along that supporting shelf, to get wholly on it, so that I might work myself erect, but after the first motion, I immediately quit moving and lay still, terror gripping me completely. All along beneath me I felt sharp points jabbing upward into my diving suit, sawtoothed steel protruding everywhere from the support beneath me. What I had come to rest on there in the mud was twisted wreckage torn from the *Paulding's* keel as she raked over the *S-4*, wreckage now sunk alongside her victim. And if I moved, those razor-edged steel plates would cut my suit wide open in a dozen places as the *Pauld-*

ing's bow had already ripped apart my glove, and drown me out of hand!

I lay still. No more struggling, no more efforts to rise. In the blackness, in the cold, in the mud 20 fathoms down, I lay quietly, not daring to move a muscle, hardly daring to breathe, desperately wondering what to do, with the one slim relief in my situation that I was resting on my right side with my cut right glove beneath me so that I could still keep air in my suit.

Then the answer came to me. Simple! No reason at all for me to worry. I should have seen it before. Inside my helmet I had a telephone; tied to my breastplate I had a lifeline. All I had to do to get clear of my trouble was to call my tenders on the surface and tell them to pull on my lifeline. They would immediately heave me up off that terrifying broken steel, up through that clinging mud which was engulfing me in maddening darkness.

Carefully, slowly, I twisted my head round to bring my mouth opposite the telephone transmitter in the roof of my helmet, fortunately clear of the water lapping round my neck, and gasped out:

"Topside there!" Then I waited with straining ears for a reply. Had the water already short-circuited my telephone? With infinite relief I heard the answer from the *Falcon* in the world above me:

"Topside. What is it?"

"Heave in on my lifeline!"

"Aye, aye. Right away!"

My head dropped back thankfully. All my troubles were over. With muscles tensed and stiffened legs, I waited for the tug on my breastplate which would pivot me about my lead-clad feet, pull me erect, and then up and off that terrible bed in the mud. But the tug never came and after what seemed to me an endless wait with not the slightest pull on my lines that I could feel, in anxious tones I sang out again:

"Topside there! Are you heaving yet?"

The answer fell like a sledge hammer blow on my strained nerves.

"Yes! Four men are heaving hard on your lines but they can't get an inch of slack. What's the matter down there?"

Four men heaving on my lines and I couldn't feel even a slight pull! And they were getting in no slack at all on deck. Then my lines must be afoul of something above me, probably tangled in the *Paulding's* broken stem which had already pierced my glove! The strain that four men heaving hard on deck could put on my air hose fouled on the sharp edges of that jagged stem projecting from the *S-4's* side could easily cut my rubber air hose in two, leave me there in the mud to be asphyxiated! In a strangled voice, I screamed:

"Topside there! Avast heaving! Slack off! Slack off for God's sake!"

An agonized moment passed while I waited in suspense. Would they get that message in time, would they obey it? Then almost with a sob of relief I heard:

"We're slacking away! Do you want any help down there?"

Did I want any help? Heaven knew I needed it badly enough, but I could waste no more breath in talking.

THE TWISTED WRECKAGE OF THE "PAULDING'S" KEEL AFTER
BEING FISHED OUT OF THE MUD ALONGSIDE THE "S-4" BY
DIVER CRILLEY (REAR)

My head sagged back, I didn't bother to answer, and I heard nothing further. On the topside, I knew that they would do what they could anyway, probably dress a relief diver and send him down; but, in my position, I gave up expecting any further aid from the *Falcon*. I must rely on myself. Long before they on the topside could do anything for me, I should probably be finished. The *Falcon* and my shipmates on her might for the next few minutes just as well be on the moon so far as helping me was concerned.

Despair, black as the blinding night around, gripped me. On top of all else now, my air hose fouled and likely to be severed, leaving me loaded down with heavy ballast to choke to death! And if I moved to extricate myself, the chances were excellent that I should cut my suit to pieces and drown in a flood of mud gushing into my helmet!

I tried to think. What could I do to save myself? But on the bottom, 20 fathoms down, with the pressure of the whole Atlantic numbing my brain, thinking was next to impossible. Then that paralyzing blackness! Light, light! If I were going to die, it would be so much easier if only I could see a little! And all the time there kept obtruding into my frantic efforts to pull my thoughts together, to think coherently if only for a moment, a shattering vision, the image of that silent, motionless steel coffin, the *S-4*, whose deck short minutes before I had been treading, with forty men stretched cold in death inside her. And there I was, the forty-first, already buried in the mud alongside their tomb, ready to join them!

Silently, motionlessly, I lay there in the ooze of the ocean floor, trying to concentrate on my problem, momentarily expecting my suit to give way to that piercing steel, to feel the mud gushing in on me. No one could do anything for me in time. What could I do to save myself? Desperately I strove to put that recurring vision of the submarine out of my mind, to concentrate on diving. What trick in diving technique that I knew or could imagine could get me safely off that broken steel beneath me, up out of the Stygian blackness of the mud enveloping me? With an effort, I canvassed every diving trick I could think of, painfully trying to fit each one to my situation, dejectedly rejecting one after another as offering no hope. The only obvious one, my lifeline, I had tried and it had failed. Nothing else seemed applicable. With body numbed from cold, nerves deadened by repeated shocks, and mind dulled by heavy pressure, I lay now in a torpor. Further struggle seemed hopeless.

Then gradually I became vaguely aware, in the utter silence and blackness of that grave beneath the ocean floor, of a persistent murmur in my helmet, of a murmur to which long familiarity had made me oblivious. A small current of air was still flowing through my helmet, escaping with a gurgle through the exhaust valve somewhere near my chin. I still had air coming through my hose from above. Suddenly across my dazed brain that brought an idea. That air could save me!

Slowly, cautiously, not to let any movement of my body saw through my suit, I dragged my left arm through the clinging mud to my breast, fumbled with frozen fingers encumbered by mittens and stiff diving glove till I found the handle of the air control valve bolted to my breastplate. Through that valve the air from the *Falcon's* compressors, coming down my air hose, entered my helmet.

Convulsively my fingers closed on the valve handle, twisted it wide open. Immediately a suddenly increased stream of compressed air roared into my helmet, started to inflate my suit. I could feel the canvas which had been pressing in on my chest ease off as if a heavy weight had been removed and begin to swell. Under the increasing buoyancy, my helmet lifted a trifle, then, as if a giant hand had seized me by the shoulders, my body started to float upward through the mud, to come erect as my suit swelled more and more. Another moment and I was free of that bed of torn steel plates, erect once more, and could feel myself dragged vertically upward through the mud by my over-inflated diving suit!

A little further and my helmet burst through the ooze of the ocean floor into the water. Light! Blessed, soul-satisfying light streaming through my faceplate again! After the terrifying darkness of the mud, that dim half-twilight of the depths seemed to me as dazzling as if the sun had suddenly risen inside my helmet! I had light now; anything was possible! I gasped in relief. I was saved!

But my dangers were not yet over. Under the pull of my partially ballooned-out rig, I was still rising from the mud, excessively buoyant. If, when I tore free of that clinging ooze, I was still so light, I would go shooting upward through the water with ever-increasing speed as the sea pressure decreased and the swelling air in my suit ballooned it out further, perhaps to crash at high speed into the *Falcon* above and kill myself, or, if I missed her, to break through the surface of the sea like a salmon leaping at a fly and then fall back helplessly, immediate victim to "the bends," having risen from the heavy pressure of the depths to the surface without the slow decompression which alone could avoid it.

No, thankful as I was over the light, I couldn't afford to take time giving thanks, or I should find myself "blowing up" from the depths. Instantly when my helmet popped out into the water, my still-buried fingers were clawing again through the mud for my control valve, shutting off the air before it was too late. When the excess air had blown itself off through the exhaust and my ballooned-out rig had shrunk back to more normal proportions so that I ceased rising, I found myself still buried to my waist in mud with only the upper half of my body in the water. But I dared not float myself any further up for fear of not stopping. And there, in equilibrium, half in water, half in mud, I stood suspended in the ocean floor.

I looked up. To my pleased surprise, my lifeline and my air hose were floating vertically above me with no tangles in sight. Were they still fouled, or had my rise from below permitted them to slack off and to come clear of trouble? I could quickly find out. Once more I called the *Falcon*.

"Topside there! Take an easy pull on my lines!"

A brief moment, and then I felt a gentle tug on my breastplate. My lines were free! Swiftly I bled more air from my rig to make sure I was heavy enough not to float and then:

"Topside there! Pull me up ten feet!"

My lines promptly tautened and up I went, legs tearing free of the mud, and when the heaving ceased, there I was, clear at last, a fathom off the bottom, dangling in the water from my lifelines. And now to get back on the submarine so that I could come up the descending line.

Where was the submarine? Slowly I kicked myself around in a complete circle, beating the water with arms and legs, but in no direction could I catch any glimpse of her. While I must be surely within ten feet of it, for I had fallen straight down its sheer side, yet the S-4 had completely vanished in that translucent water.

I gave up all hope of having the tenders above land me on the sub; even more out of question was finding her myself by having them lower me to the bottom again and floundering around in that mud.

My inspection was done anyway—by a freak of fortune, successfully. My hour on the bottom was certainly up. A few more words over the telephone and I was started directly upward for the surface, on my slow and tedious rise through the icy sea, a rise broken by lengthening stops every few fathoms as the water pressure decreased, to allow the air to work itself out of my blood gradually, to decompress me and avoid "the bends."

And so, step by step, I was lifted. Halfway up I clambered aboard the little steel stage lowered into the sea alongside me. Then up again till waving over my helmet I could once more see through my faceplate that foam-flecked undulating sheet, the surface, with the familiar red underwater hull of the *Falcon* nestling in it. One more last stop, and then at last the welcome message:

"Coming aboard!"

With a final heave, the stage rose and I burst through the surface, clinging tightly to the bails of the stage again to avoid collapse as my buoyancy vanished and the unsupported load of all my lead and copper ballast came suddenly down on my shoulders. Swaying violently, the stage rose over the bulwark, swung inboard, dropped with a bang on the deck. Dripping mud and water, I was seized by the tenders and dragged to a bench. My belt, shoes, and helmet were hastily stripped off, and then, without a pause, still clad in my dripping suit, I was rushed across the deck, up the passage and unceremoniously jammed through the outer door of the recompression chamber into the first lock. Slam! went the door. A tender with me twisted open an air valve and in roared a stream of compressed air, once more to get me under pressure, to make sure that no bubbles of air formed in my veins to give me "the bends."

Exhausted, I sank down on the deck. As the pressure rose, the tenders in the lock with me unbolted my breastplate and dragged me out of my mud-plastered suit. By the time that was done, the pressure had risen to balance the pressure in the inner chamber, and the round door to that swung back. In the inner lock, I would finish my decompression, surrounded by hot towels and hot drinks. Chilled and shivering, I crawled through the round door into the inner chamber.

To my surprise, I found I was not alone in that recompression tank. Stretched out on the floor of the inner chamber, clad like myself in blue diving underwear, was Bill Carr, while alongside him was Tom Eadie, rubbing his legs. I looked at them inquisitively. Carr, with his underwear half soaked and his legs blue with cold, had evidently like myself recently come up from a dive.

"Queer," I thought. "I didn't know any other diver was on the bottom since I went down."

"Been down, Carr?" I asked him.

"Yeh," answered Carr briefly, evidently too chilled to want to talk.

Eadie hospitably offered me a drink, one of our hot "submarine cocktails." Greedily I swallowed it, thankful for the fiery warmth that went racing through my frozen form. Eadie, as I drank, once more turned to Carr, kneading his arm muscles to warm them up. I sank down on a bench. Eadie, still working, looked up at me.

"Didn't you know Carr was down, commander?"

"No, Tom," I muttered wearily. "I thought I was alone down on the bottom."

"Well, you weren't," said Eadie. "You hadn't been down five minutes when they sent Bill down with a fire hose to start washing a tunnel through under the port bow of that sub, so's we could get the lifting chains through under her there. You didn't see nothing of him?"

"No, Tom," I assured him. "Not a sign."

"Well, he was there," continued Eadie, his gray eyes fixed solemnly on me the while he vigorously massaged Carr. "And he was down there when you telephoned up first to heave on your lines and then yelled to quit heaving and we knew you was in trouble. So we telephoned down to him, seeing as he was already on the bottom, saying:

" 'Commander Ellsberg's in bad trouble somewhere on the sub. Leave what you're doing, Bill, and go over and help him!' "

I looked from Eadie to Carr, who, still stretched out at my feet, was apparently vigorously trying to signal to Eadie to shut up. But Eadie, ignoring him, went on:

"And d'ye know, commander, what Bill said when we told him you were in trouble and to go over and help you?"

"No, Tom," I replied, "that's beyond me. What did Bill say?"

"Well," said Eadie, "he sings out into his telephone:

" 'Aw, tell him to go to hell! I'm stuck in the mud myself!' "

This was too much for the prostrate bosun's mate. Carr came instantly to a sitting position, almost knocking Eadie over backward in the flurry. His face, flaming red, shot one scornful look at Eadie, then turned toward mine.

"Say, commander! Don't believe a word Tom says! You know I'd never say a thing like that about you! You wanna know what I really said?"

Courtesy Commander Henry Hartley, U.S.N.

THE "S-4" IN DRY DOCK IMMEDIATELY AFTER BEING RAISED

"Sure, Bill, don't worry. Tom's only kidding. I know you'd never throw a shipmate down like that. What did you say?"

"Well," said Carr, "there I was alongside the port bow, washin' away with the hose, when the topside phones me about you. Quick as a wink, I drops my hose, grabs my descending line, starts climbing up the port side o' that sub, thinkin' o' the argument you once gave me in New York over that letter o' commendation, because I'd never had a chance to be a hero. And when I gits on deck, I starts runnin' aft four bells lookin' for you an' shoutin' into my telephone:

" 'Hooray! Here's where I gits that Navy Cross!' "

Carr paused, looked mournfully at me, then concluded sadly:

"Aw, commander, why didn't you wait till I got there to rescue you?"

CHAPTER VI

WHAT happened to the *S-4?* I made my report to Captain King, carefully outlining the condition of the submarine and the lack of structural damage, and salvage operations began.

By now, a few days before Christmas, our salvage gear had all arrived, and so also had practically all the divers used in the previous salvage operations. A major problem to decide was whether to wait for spring and better weather for salvage, or to work on through the winter. Based on the *S-51* experience, the former seemed preferable, as on that location 14 miles at sea off Block Island, a succession of gales had made winter work impossible. At Provincetown, however, the tip of Cape Cod, directly to the eastward, offered protection against easterly gales. Only against northwesters, like the one which tragically enough blew during those fatal first days, was there no protection.

While this matter was being discussed, we had a visit from the Secretary of the Navy, Curtis D. Wilbur, who settled the matter for us. Regardless of the weather, the job was to proceed, he ordered. And so it did.

Rescue operations were discontinued; the forces present were reorganized on a salvage basis only. Admiral Brumby left to resume command of his forces afloat, Captain King was put in complete command, and I, with nothing further possible in the way of rescue, said good-by to my shipmates and, on New Year's Day, steamed back across Massachusetts Bay in a dead calm, bound for home. Commander Saunders became Salvage Officer, the job I had had on the *S-51* operations, and Lieutenant Hartley, skipper of the *Falcon,* took general charge of the divers.

Luck was with them. Fate, as if satisfied with what it had already dealt in the way of death and disaster, sent no more storms. Only the cold and the cold water remained as unusual obstacles. Fighting these, the divers clung to the task. Using the same equipment and the same general methods by which, two years before, the *S-51* had been raised, they sealed up the inside of the submarine, expelled the flood waters from the undamaged compartments, tunneled under to pass lifting chains and sank and attached pontoons. Finally, on March 17, 1928, three months to a day from the time the *Paulding* sank her, the *S-4* rose from the depths, burst through the surface, and was towed to the dry-dock in Boston.

Seemingly, Fate dogged the *S-4.* First, in the unfortunate combination of circumstances, never wholly explained, which brought her up in the *Paulding's* path; next, in the fouling of the ventilation valve which drove her crew from the control room; and last, in the occurrence of the accident just before a three-day gale, the worst of the whole winter, which, except for the initial dives, held the rescuers off till it was too late to save a single life. Now, twelve years later, I am still convinced that, with the men and means then at hand, except for that long-continued gale, we would have saved Fitch and his five companions in the torpedo room. We might even have revived and finally have rescued at least the stronger members, if not all, of Jones's party in the stern.

Except for that long-continued gale—

There is but one consolation. In the future we shall not have to make that exception. As in every great disaster, the deaths of the *S-4's* crew drove home vividly to the nation, as no accident before had ever done, the perils and the needs of the submarine service. Money and men were for the first time freely made available by Congress to allow experimentation and improvement— more rescue ships were provided to cover more closely the areas in which submarines work, more pontoons, more divers, improvements in the submarines themselves. But most important of all, a submarine escape device, originally conceived abroad, was, with the funds provided and in the hands of the captain of one of the *S-4's* sister submarines, developed into a usable appliance, the Momsen "lung." With this apparatus, now issued to every submarine, the crew may escape from any compartment of the boat and rise to the surface, wholly without aid from other vessels, wholly independent of the state of the weather.

There has been one other peculiar result which psychologists may well ponder. There has never been a modern submarine accident that was not in some degree the result of a lapse in vigilance by some member of the crew. In our Navy, from the time we first had workable submarines up to 1927, we had a serious submarine accident, resulting in a sunken boat, at least once every two years, though in most cases part or all of the crew were saved. Perhaps men cannot live in a state of eternal vigilance, but such was the impression made on the minds of other submarine crews by the deaths of their shipmates on the *S-4* that twelve years have since gone by without the sinking of a single submarine, whereas on the basis of past performances in the submarine flotilla, six boats should have sunk in that period.

Evidently in the memories at least of their shipmates of the submarine service, those hammer taps are still heard as vividly as on the sad days just before Christmas, 1927, when they rang out on the bottom from the torpedo room of the *S-4*.

CHAPTER VII

PUBLIC wrath sometimes serves a useful purpose. The indignation of the country as a whole over the deaths of the *S-4's* crew, trapped inside their own boat on the bottom of the sea and unable to escape, centered attention on the problem in the one spot where attention was worth something, in Congress. No submarine crew will ever again face the situation that the men inside the *S-4* did, not because anything startlingly new has been invented since, but mainly because Congress provided the money to take previously existing ideas and make realities of them.

A long time before the *S-4* was built, as far back as 1903 when submarines were hardly more than interesting toys, the first submarine escape "lung" was devised in England by R. H. Davis, manager of the firm which long before had produced the modern and now universally used flexible diving dress. The original "lung" was crude, to some degree complicated, and naturally imperfect, but it had considerable possibilities and would unquestionably have been useful in saving life had it been available in accident cases on submarines. However, neither the Davis "lung" nor any substitute was installed on our boats; the men inside the *S-4* had no escape device of any nature to permit them to emerge before they suffocated, though at that time apparatus for such purposes was at least twenty-four years old.

The public clamor over the S-4 put a different complexion on the matter, both at home and abroad. The "lung" idea as an escape device was taken up in earnest by many navies, and, as usually happens when money is available for experimentation, development of neglected ideas is very rapid. In our Navy, Lieutenant C. B. Momsen took the "lung" idea and, assisted by technicians in the Navy Department, soon turned out a much simplified device which he proved to be workable by using it himself to escape from submarines at depths greater than that at which the S-4 had bottomed. Abroad, the British Navy, which had also suffered a number of submarine disasters contemporaneously with ours, began to take the Davis "lung" seriously and shortly adopted an improved model for its submarines.

"Lungs" are simple devices to enable a man to breathe under water for a period while rising from a sunken submarine. They resemble gas masks in appearance, being intended to permit a man to continue breathing safely when immersed in an irrespirable or poisonous atmosphere, the main difference being that for the user of the "lung" the surrounding atmosphere is a liquid instead of a gas.

The essential feature of all "lungs" is a bag of approximately the capacity of a man's own lungs, into which he exhales and from which he inhales, using the same air over and over again, but each time passing the air exhaled through a cartridge of chemicals that remove from it the carbon dioxide which otherwise would soon make the contents of the "lung" dangerous to life.

The Davis "lung," in all its forms, has always carried in addition to the chemical cartridge a small oxygen cylinder, from which fresh oxygen can be continuously supplied to replace that used up in breathing, so that the wearer can rely on it for upwards of an hour and a half. The Momsen device omits this oxygen flask and relies on charging the bag itself with oxygen when first put on, so that it is usable for a much shorter time, though long enough to permit escape.

In the use of any "lung," it is necessary for the men caught in a sunken submarine first to flood the compartment in which they are, until the pressure inside rises to equal the sea pressure outside, so that a hatch can be opened to the sea. What air there is inside the submarine is, during this period of flooding, compressed into the upper part of the room while the water rises, and is trapped behind a screen projecting downward from the near-by hatch so that, when the hatch finally opens to the sea, the men are left with their heads in a large bubble of compressed air held in the top of the compartment while their bodies are probably immersed in the water.

At this stage, with the Momsen "lung," each man inflates his bag with oxygen from a large flask in the room, puts a clip over his nose, slips the mouthpiece of the apparatus into his mouth, gets a good grip on it with his teeth, and starts breathing into his "lung." Then from the inside of the room, a buoy on a long line is released and paid out till it reached the surface, when the lower end is secured inside the submarine. The first man to go

From "Deep Diving"; Courtesy Sir Robert H. Davis

RISING THROUGH THE SEA WITH THE DAVIS ESCAPE APPARATUS

out then ducks under the screen inside the submarine and goes up the hatch into the water. He seizes the buoy line and rises on that, pausing every 10 feet and breathing continuously into his "lung." The others follow one at a time.

At the surface, the "lung" has buoyancy enough to act as a life preserver and keep a man afloat till he can be picked up. The ideal condition, of course, is for the men below to wait until they know a vessel is above to rescue them, even though it be only a rowboat; in desperate cases, the men might have to emerge, to avoid death below, before they were sure there was anybody to pick them up, and take their chances as swimmers in the sea.

It was made painfully evident soon after "lung" devices were issued to our submarines and training in their use started, that it was quite as dangerous to put a "lung" into the hands of an untrained seaman and tell him to use it in rising through the water, as it was to plank a man, ignorant of how to drive an automobile, down behind the wheel of a machine going 60 miles an hour on a crowded highway. Two deaths and a dozen serious cases of mysterious illness shortly brought out the fact that an untrained man in a "lung" could kill himself quite promptly, though what was doing it was not immediately clear.

But to see a man, who in practicing with a "lung" had risen through only 16 feet of water to the surface, come up apparently all right, take one or two vigorous strokes in swimming toward a boat, and then suddenly collapse and die before he could be hauled aboard, was disconcerting, to say the least. After that had happened in our Navy twice, even to men who had been instructed in the use of "lungs," a long medical investigation was carried out, ending with a careful series of experiments on animals in the Harvard Medical School. This demonstrated conclusively the cause and the cure.

The cause was the natural reaction of a man under water, when he is frightened, in holding his breath in an involuntary attempt to avoid swallowing or breathing water. But it turned out that there is no surer way for a man to kill himself promptly when using a "lung" than to do this apparently natural thing. For a man rising from the bottom starts with his lungs full of air at a pressure corresponding to the water pressure on the bottom. As he rises, the water pressure decreases. If he keeps on breathing naturally as he goes up, the pressure in his lungs keeps decreasing to correspond, with the excess air escaping continuously from the "lung" through an automatic exhaust valve.

If, however, he becomes frightened or hysterical at finding himself completely immersed in the ocean and perhaps a long way from the surface (a mental condition not unnatural in a person not trained as a diver) and does what seems indicated as the normal thing, that is, holds his breath, he practically seals his death warrant. For as he rises further, probably faster now in his haste to get to the surface, the water pressure surrounding his chest decreases, but the air pressure in his lungs, with breathing suspended, does not. The result is that the air trapped in his lungs expands and in so doing distends the lungs abnormally. It takes a rise of not

over six feet in the water under such circumstances for
the distention of the lungs to cause myriad ruptures in
their walls through which air in considerable quantity
is forced directly into the blood in large bubbles.

Nothing further happens until about a minute after
breathing is resumed, when reaction is swift. The air
bubbles then reach the heart, the brain, the nervous
centers, and death results almost immediately from air
embolism.

So all "lung" training concentrates now on insisting
that the user, whatever happens, keep on breathing continuously while he rises. To insure this, our Navy has
built several training tanks for giving submarine crews
experience with "lungs" under practically service conditions. In these a rise through 100 feet of water can be
made, with intermediate locks for training men at 18-
foot and at 50-foot depths. No man is now permitted
as a member of a submarine crew who has not been
trained in the use of the "lung" and who has not been
qualified by an actual ascent in a tank, demonstrating
his knowledge of proper escape technique.

The British Navy, as mentioned before, since 1930
has used the Davis "lung," and to train its submarine
forces has two tanks, one at Hongkong and one at
Portsmouth. So far, the Davis device is the only one
which has been tested in an actual wreck. It was issued
for general use to British submarines in 1930. In the
early summer of 1931, on the China station, the British
submarine *Poseidon*, which had just come up from a
dive, was rammed by the Chinese steamer *Yuta*, and

sank immediately. The major part of the crew (some of whom were already on deck) escaped from the conning tower before the *Poseidon* disappeared and were picked up by the *Yuta*, but 18 men went down inside the submarine.

Of these men, 12 in the after end were never heard from, and undoubtedly drowned immediately as the ship flooded aft. But forward in the torpedo room were 6 men, under the command of a torpedo gunner's mate, P. H. Willis. Willis succeeded in closing the torpedo room door against the inrushing water as the *Poseidon* went down, shutting himself and 5 companions into the torpedo room, and there then was almost exactly duplicated in the *Poseidon* the situation in the *S-4* four years before.

Once more a submarine was sunk in about 20 fathoms of water; once more 6 men were trapped in her torpedo room; once more there were vessels on the surface unable to do anything immediately to help those below. But there was one difference—the men in the *Poseidon's* torpedo room were equipped with the Davis "lung" and, under Willis' leadership, promptly set out to help themselves.

While Willis was flooding the torpedo room, one of his seamen, Nagle, undertook to instruct the Chinese steward, Ah Hai, who was caught forward, in the use of the "lung." The others in the room had already had some training.

After some two hours on the bottom with water flooding in, the water had risen high enough so that Willis concluded the pressure had equalized and they might try opening the hatch. With some difficulty, the

U. S. NAVY SUBMARINE ESCAPE TRAINING TANK AT NEW
LONDON, CONNECTICUT

upper hatch was opened, and two seamen, Lovock and Holt, each wearing the "lung," passed through and started up, but the inward flow of water slammed the hatch to again, and the remaining four men could not get out.

Lovock and Holt rose to the surface through 120 feet of water, but when they got there, Lovock was unconscious and died immediately (a victim probably of air embolism due to a failure to keep breathing while rising). Holt, seeing only that he was unconscious, supported him till both were picked up by boats hovering over the wreck.

Meanwhile, inside the partly flooded torpedo room in the *Poseidon*, Willis and his three shipmates were worse off than ever. A considerable amount of the air inside the room had blown out of the hatch when it was first opened; they had now to admit far more water to compress what air was left enough to allow opening the hatch again. So for another hour they had to stand silently by while they finished flooding, by which time they were in water up to their necks and very little air space was left above them for breathing.

Willis now tried the hatch again, and found that it would open. Four hours had gone by since the *Poseidon* sank, when one at a time the last four men (including the Chinese steward) floated up from the depths to be picked up by the boats, all safe.

There on the *Poseidon* in 1931 the "lung" proved its value. And Robert H. Davis, who twenty-eight years before had turned out the first "lung," was knighted by his king, George V, in recognition of his services in providing a practical way out for trapped submarine sailors.

CHAPTER VIII

DIVING is queer business. Dangerous? Yes. But not for the weird reasons that lurid fiction writers and melodramatic authors, who usually know no more of diving than they do of conditions on Mars, have fed their readers. Why didn't the divers on the *S-4* save the men in her torpedo room; why did they make only a few dives before the storm drove them off; why didn't they keep on diving, storm or no storm; why when they did dive did each man accomplish so little?

From the thousands and thousands of protests that poured in on Washington and the flood of press criticisms that descended on the rescue fleet at Provincetown, it was evident that the people in this country as a whole knew as little about diving as did Victor Hugo and Jules Verne who long ago fixed in the public mind a fantastic picture of divers and diving which has never since been eliminated.

What makes diving difficult and dangerous? To the diver the world over, whether Greek, Kanaka, Japanese, Italian, or American, who is diving to accomplish a specific commercial result and not merely to provide faked movie thrills or melodramatic books, it's neither the devilfish nor the shark that must be battled hand to hand in desperate combat on the sea floor. It isn't the villainous conspirator bent on severing a diver's air hose and robbing him of the treasure he has salvaged from the deep—those chests of gold and bushels of pearls and rubies which the public has been led to believe always pave the bottom of the sea. It is none of these fantastic imaginings. What the difficulty actually is—what the danger always has been from the days, thousands of years ago, when the first savage, plunging beneath the surface, sought to bring up sponge or pearl from the depths—is the very prosaic fact that water has weight, plenty of it.

As nature made both us and this world in which we exist, we live and have our being at the bottom of an ocean of air—a very deep ocean, extending scores and scores of miles above us, but with the air rapidly thinning away toward a vacuum only a few miles up. The cumulative weight of this gaseous ocean presses down on us, enters our lungs, permeates our bodies; and nature has through countless eons evolved our organisms to stand this pressure which, registered on the ordinary barometer, averages about 14.7 pounds to the square inch. This we call a pressure of one atmosphere. Our brains, our nerves, our lungs, our entire bodies are accustomed to function under this atmospheric pressure; if it be varied much, either up or down, discomfort and even distress promptly follow.

It is a widely observed fact that people with weak hearts find it dangerous to cross high mountains, to live in high altitudes, or to travel very high in airplanes. The reason, of course, is that as such individuals go up and the air pressure decreases they commence to pant violently, trying to get enough of the rarer air into their lungs to give their bodies the oxygen they need. Their weak hearts sometimes give way under the added strain.

Yet even under these circumstances, the change of pressure and variation in the amount of air inspired per breath are relatively insignificant compared with the changes which a deep sea diver has to bear. For instance, atop Pike's Peak, some 14,100 feet above sea level, which is as high as most people are ever likely to get, the pressure drops about four tenths of an atmosphere—from 14.7 pounds per square inch (which is the sea level average) to around 8.7 pounds, decreasing some 6 pounds. This decrease of about 40% from normal in pressure causes many people acute distress and makes even the average person pant violently in climbing the few steps to the observatory tower, so closely has nature tuned us to what she considers normal and so little are we able to bear deviations from that normal.

Consider now the diver. Sea water is heavy; a cubic foot of it weighs 64 pounds. For every foot below the surface the diver goes he has added to the weight of the sea pressing on each square foot of his body a further load of 64 pounds; or .445 pounds on every square inch of his body. He need descend but 14 feet below the surface of the sea to raise the pressure on his body by the same amount, 6 pounds, that an ascent of 14,100 feet to the top of Pike's Peak decreases it. And he need go but 33 feet below the surface of the sea to double the pressure on him—to make it twice what nature intended him to bear.

From "Deep Diving"; Courtesy Sir Robert H. Davis

LATE TYPE ESCAPE CHAMBER AS FITTED ON MODERN BRITISH
SUBMARINES

Now 33 feet is no great depth for a diver who must take wrecks where he finds them; working depths beyond 100 feet are not uncommon. What does that mean to the diver?

On the S-51, we worked for months at 132 feet, 22 fathoms down. At that depth, the water pressure on the sea floor was 4¼ tons to the square foot, or 59 pounds to the square inch, and the diver had a total load bearing on every square inch of him of 5 atmospheres, just five times what nature had designed him to stand! Over the entire surface of his body, lay a pressure of nearly 60 tons, a load easily capable of reducing him to jelly (and which to some unfortunate divers has done just that).

Owing then to the extreme weight of water, it is obvious that a deep sea diver faces extraordinary conditions and dangers under which he must live and work. Compared to these perilous conditions, the tinsel devilfish and sharks with which the literature of diving is festooned fade into insignificance.

What results? To live and work under water for more than a few scant minutes, it is of course obvious that a diver must continuously be supplied with air. To prevent his chest from being crushed in by the terrific load on it in deep water, this air must be supplied him at a pressure slightly greater than that of the water surrounding him. And mechanically to permit him to breathe at all, there must be some device, over his head at least, to keep the water away from his nostrils while he breathes.

Centuries of slow development evolved at last, in

the hands of Siebe, Gorman and Company of London, the present combination of a rigid copper helmet bolted tightly to a watertight canvas and rubber suit, properly weighted, as best fulfilling the mechanical necessities of a diving rig. Air is fed into the helmet in a steady stream, and, after it has been breathed, escapes to the sea through a spring-loaded exhaust valve. The air breathed into the lungs, at a pressure slightly above that of the sea bottom, communicates its pressure through the blood to every part of the diver's body, putting him under an internal pressure counterbalancing that of the sea pressing on him externally. In this condition of equilibrium, the diver can stand any depth of sea without being crushed.

Normally then the diver is in balance between two opposing forces of great magnitude, and his state may fairly well be illustrated by comparison with that of a pneumatic tire on a heavily loaded truck. So long as the tire is inflated with sufficient compressed air, it stays rounded out and supports the load pressing down upon it. But if the tire "blows out," down comes the weight of the truck upon it and flattens it to a pancake. In the same way, if the diver by any chance loses the air pressure in his suit, down comes the weight of the sea upon him and instantly crushes him as flat as any blown-out tire.

To live then beneath the sea, the diver in the usual helmet and flexible suit must breathe compressed air, under pressures of three, four, five, or even more atmospheres, pressures far above that existing on the surface, to which nature has accustomed us; and the re-

sults of breathing under such unnatural conditions lead
directly to the gravest danger that a diver faces, that
of "the bends."

Not many decades ago, when diving was relatively in
its infancy, though the diving rig used had been fairly
well developed to its present state, divers began to notice that, if they went much below depths of 60 feet
or stayed down very long at that depth, say much
over an hour, when they came back to the surface they
were shortly attacked by a mysterious disease, causing
intense pain and doubling the victim up in strange con-
tortions. From the convulsive movements of the sufferers the malady received the slang name "the bends."
For those unfortunates who had gone much below 60
feet or had worked there several hours, "the bends"
often took a more serious form; paralysis set in, making
the victim a helpless cripple for life, and in some cases,
even, quick death ensued, coming within a short time
after emerging from the water.
   What caused these afflictions the divers could never
figure out. It was simply obvious that strange death and
quick disaster lurked in the ocean depths for such as
dared penetrate them, usually in search of sunken treasure. What is unaccountable is naturally doubly terrifying, and to the men involved it seemed that malignant
demons lurked on the ocean floor, intent on punishing
the plunderers who sought to rob the wrecks the sea
had once claimed for its own. Still the lure of sunken
gold is strong; some men escaped attack, and, in spite
of sudden death and of even more horrible complete

paralysis, deep diving continued spasmodically, with the victims of "the bends" fatalistically accepting their curse, no more able to explain what had struck them, nor why, than we with all our boasted modern science can illuminate either the cause of infantile paralysis or the erratic manner in which it seems to select its victims.

Had it been left to divers only to solve the mystery, it is probable that the matter would still be in the realm of superstition, but the advancing needs of engineering brought the problem forcibly to the attention of scientists. The construction of tunnels under rivers (developed in the nineteenth century) and the necessity of providing deep foundations along river banks for huge bridges brought in as an indispensable tool the caisson. This, maintained under moderate air pressure, kept back the water while the laborers, euphemistically called "sand-hogs," burrowed through the water-soaked muck and clay to advance their bore or sink their footings. Here again were soon noted the symptoms of the dreaded diver's malady; but now, with hundreds of men involved, the fatalities jumped, the number of cases of "the bends" multiplied, and the work suffered.

But "the bends" did not confine itself to the sand-hogs. It struck at a more notable victim. The Brooklyn Bridge, with by far the longest span of its time, required extraordinary caisson work for its pier foundations in the soft mud beneath the level of the East River. Its builder, Washington Augustus Roebling, spent so much time personally supervising the caisson work on the success of which his mighty bridge structure depended that he shortly was struck down himself by the compressed air illness, and for ten years, from 1873 to the completion of the bridge in 1883, was compelled to direct the construction from his sick bed.

"The bends" began to receive scientific attention for the first time. Under the more dignified title of "caisson disease," science turned an inquiring eye on the problem, and very quickly dissipated the mists of superstition. In 1860, as a preliminary, came the first inkling, when a Frenchman, Professor Leroy de Mericourt, expounded the theory that the diver's blood becomes surcharged with the air which he breathes under heavy pressure; but it was left to his compatriot, Monsieur Paul Bert, in 1878, after Roebling's disaster, first to investigate the disease seriously, to expound the causes of "the bends," and to indicate the first crude method of avoidance.

Bert demonstrated that the disease was due wholly to bubbles of a gas, nitrogen, which appeared in the blood and tissues of a diver on emerging from the surface. Professor de Mericourt had demonstrated how the gas came on the scene. Briefly the situation is explainable about as follows:

The air we breathe is composed roughly of 21 per cent of oxygen (the essential gas for sustaining life), and 78 per cent of nitrogen (a wholly inert gas which simply dilutes the oxygen), with the remaining 1 per cent a mixture of carbon dioxide and certain other rare gases which may here be neglected. As we breathe under ordinary conditions, this mixture of gases is inhaled into the lungs, where, over the extended surfaces and cells of those organs, it comes into intimate contact with the blood. The result is that part of the oxygen is converted into carbon dioxide and the remainder— together with the nitrogen, the newly formed carbon dioxide, and some additional moisture—is expelled on exhalation.

But the diver's conditions of breathing are far from ordinary. If, as on the S-51, he is breathing air under a pressure of five atmospheres, a strange thing happens with each breath. So great is the pressure now in his lungs that the nitrogen, instead of passing harmlessly out, is forced to dissolve in the blood. With each repeated breath more nitrogen is dissolved by the blood stream it meets in the lungs, to be carried thence to every part of the body where, under the same superpressure, it is taken up by the tissues, the fats, and the muscles.

The speed with which this absorption of nitrogen goes on is proportional to the pressure under which the diver is working, and its extent depends on the length of time he remains under that pressure. Consequently, deeper water means faster absorption and a longer stay means more gas dissolved, the total amount absorbed being a combination of both factors—depth and time.

Now while the diver is on the bottom, he never feels this process going on and is usually quite ignorant of it. But when he starts to ascend, trouble starts. In the old days it culminated in "the bends." As the diver rises, the sea pressure decreases; when he reaches the surface, it vanishes altogether. But it was this excess pressure which originally forced nitrogen to dissolve in his body, and only under that pressure will the nitrogen stay dissolved there. When the pressure decreases, the nitrogen, no longer able to stay in solution, starts to appear in the blood and elsewhere in the form of bubbles. As the pressure decreases further, more nitrogen comes out of solution, and meanwhile the original bubbles expand. The diver's blood stream, instead of being a liquid, commences to become a froth.

This condition can be best illustrated by a familiar example (an example first suggested by de Mericourt). Consider a bottle of some charged water, such as ginger ale, with the original cap intact and the bottle sealed. Held up to the light, the liquid in the bottle appears as a clear, solid fluid. That liquid contains, however, a considerable amount of a gas (in this case carbon dioxide) dissolved in it under several atmospheres of pressure; but, being in solution, the gas is invisible. Pull the cap and watch what happens.

Immediately the cap is removed, the pressure on the ginger ale is removed and drops to atmospheric. With the removal of that excess pressure, the gas begins to come out of solution, the liquid is permeated with visible bubbles, and (especially if the bottle is not chilled) the ginger ale foams up and overflows the neck of the bottle in a violent froth of escaping bubbles. Even for a considerable period after the cap is pulled, gas will continue to be liberated from solution, rising in a steady stream of bubbles toward the surface.

What happens to the uncapped bottle of ginger ale happens to the unfortunate diver who rises too rapidly to the surface after a long, deep dive. But in the case of the diver, the results are far more serious. If enough nitrogen has been absorbed in his system, bubbles large enough to clog his veins and block off circulation form here or there, causing the intense pain and convulsions of "the bends." In some cases, enough bubbles are carried to the heart to fill one side of it with accumulations of gas which stop its pumping action, causing quick death. In other cases (and relatively frequent ones in the old days) the bubbles lodge in the spinal column, causing paralysis (usually of the legs). The favorite lodging spot of bubbles in minor cases is in the joints, such as the knees, the elbows, and the fingers, causing marked pain but no fatalities.

The discovery of the cause of caisson disease indicated the remedy for it. Monsieur Bert deduced from his experiments and discoveries that if, instead of coming immediately to the surface after a dive, thus quickly losing all the pressure on his body, the diver were to rise slowly but steadily to the surface, the nitrogen dissolved in his body would slowly come out of solution as the pressure decreased, in quantities not sufficient at any time to cause sizable bubbles, and escape continuously through the lungs—a process which was called "decompression."

Until 1907, Bert's method of slow uniform decompression was generally used by divers and caisson workers, very considerably reducing the number of cases of "the bends" but unfortunately not completely eliminating them. On tunnel work in New York around the turn of this century there were still numerous cases of sand-hogs, hours after they had received their usual decompression and left the job, collapsing on the streets. Ordinarily they were considered by passers-by and even police as simply sodden drunks and left in the gutters or carted unconscious to a police station to sleep off a supposed jag, which naturally they never did. After some sad experiences with cases of "the bends" thus handled, caisson workers were by law required always to wear a metal tag, warning observers that, in case of collapse anywhere, the wearer was to be rushed immediately to a specified hospital equipped to treat "the bends."

Obviously, Bert's decompression methods left much to be desired. In 1906, the British Admiralty undertook a further study of the subject, mainly carried out by Professor J. S. Haldane, a physiologist, together with the then Lieutenant G. C. C. Damant, R.N. Haldane discovered that Bert's uniform slow decompression did not positively insure desaturation of the blood and tissues. Paradoxically, Haldane's experiments showed it was much safer to decompress in stages—that is, to come up sharply part way, so as to cut the pressure on the diver in half. This sudden decrease would allow a considerable amount of nitrogen to emerge and escape, but under a residual pressure still large enough to prevent the formation of any bubbles of troublesome size. After a specified time at the first stage, the diver was to be raised sharply to the next stage, again halving the pressure on him, and here letting him stop a further period in the water for more nitrogen elimination. And so on to the surface, the idea being that each sharp decrease in pressure would bring about definite elimination, but that no one drop in pressure should be great enough to allow large bubbles to form.

Haldane's theories were correct. He worked out permissible lengths of dives to various depths, and decompression tables showing the stopping points and times of stop at each point for safe decompression. Since Haldane's day, careful observation of his decompression tables (slightly modified by later experimental work) has usually avoided attacks of "the bends." But the major drawback resulting from his work was his conclusion that, for safe decompression, the time a diver could spend on the bottom without soaking up too much nitrogen was decidedly limited. For instance, at 22 fathoms he recommended a dive not exceeding 30 minutes, to be followed by a decompression time of 33 minutes in ascending. At this depth on the S-51, owing to scarcity of divers and other reasons, we usually lengthened the time of a dive to one hour, which required a decompression time of about 90 minutes, but we were not completely free of minor cases of "the bends" and had one decidedly desperate case.

The one weak point in Haldane's stage decompression is that it must of necessity fit the normal man decompressing under average physiological conditions. Unfortunately some divers, regardless of how carefully selected, vary from the normal, and even a normal individual has his days when his bodily reactions are far from normal. Heaven only knows what may happen then. For instance, take the case just referred to.

We were salvaging the S-51, 22 fathoms down, on

the sea floor some 14 miles to the eastward of Block
Island. It was mid-November, winter was approaching, the water was very cold. Of 10 divers with whom
we had started, about 6 were knocked out by general
exhaustion and a succession of minor cases of "the
bends." If work was to continue, some fresh divers
would have to be employed.

Among the seamen acting as tenders was a petty officer, L'Heureux. He had done considerable diving in
shallow water in previous years, and only the year before, in examining another wreck, had made at least
one dive to a depth of about 190 feet. L'Heureux volun-
teered to dive on the S-51, and, as a decidedly promising candidate, was accepted.

Not to make things too difficult for his initial dive,
it was decided to send him down first simply as a helper
to another diver, Joe Eiben, who was thoroughly ac-
quainted with the wreck. L'Heureux's task was to be
only to hold a submarine lamp to light up the black
interior of the S-51's engine room, the while his mate
did the actual work of closing some valves.

Joe Eiben was dressed first, hoisted over the side and
disappeared down the descending line. L'Heureux's
helmet was screwed on, he was lifted overboard,
dropped into the sea, given the lamp, a powerful 1,000-
watt submarine searchlight, and in his turn slid down
the descending line (which was tied to the submarine's
gun just forward of the conning tower). In a little over
a minute after leaving the surface, L'Heureux's telephone tender announced:

"L'Heureux reports, 'On the bottom!' " So far, so

good. L'Heureux's tender started to pay out his lifelines, to give him slack enough to go aft astern of the
conning tower and follow Eiben down into the *S-51's*
machinery hatch.

A few minutes later, a third diver, Tom Eadie, was
dressed and dropped overboard. His task was to work
by himself on an independent job on the deck of the
submarine forward.

Ten minutes went by. Below in our hull, anxiously
watched by the engineers, compressors were throbbing,
shaking the *Falcon* rhythmically with their pulsations as
they hammered the air down through the sea to our
three divers. At the rail, three tenders periodically
"fished" the diver's air hoses to keep out a proper
amount of slack. In the *Falcon's* superstructure perched
three other tenders, headsets strapped over their ears,
transmitters in their hands, listening intently for any
message from their divers below. Around us rolled the
open sea, a moderate chop beating against the sides of
the *Falcon*, while from her, radiating in all directions
like the spokes of a huge wheel from the hub, ran out
six heavy manila hawsers, shackled to buoys in a wide
ring surrounding us to hold us steady over the wreck
regardless of how wind and sea might shift.

The calm on the *Falcon's* deck was broken by a call
from her superstructure.

"Joe Eiben's just phoned! He wants to know when
L'Heureux's coming down with that light! He says it's
pitch black inside the engine room where he is and until
he gets a light he can't do a damned thing!"

Puzzled, I looked up at the telephone tenders.

L'Heureux, of course, was on the bottom and had already had far more than enough time to get aft to his job. Signaling the tender on L'Heureux's phone, I ordered briefly:

"Tell L'Heureux to quit wasting time and get aft with Joe!"

"Aye, aye, sir!" The tender bent over his transmitter, sang out the message to L'Heureux below, then repeated it, but in another moment reported:

"Sorry, sir, but L'Heureux don't answer!"

"Give me that telephone!" I reached up, caught the headset and transmitter dropped to me from the boatskids above, swiftly slipped the receivers over my own ears and sang out:

"Hello, L'Heureux!"

No answer, unless what sounded like a somewhat explosive "Ha!" could be taken as one. Again and again I tried, but to each call, each question about where he was, why he didn't get aft, I got nothing more than that occasional "Ha!" exploding in my ears. I was getting nowhere, time was flying, the precious minutes in which Eiben should be working were being wasted while he waited in the black engine room for his light.

In disgust I sang out to the tender on Eiben's telephone set:

"Tell Joe to come up out of the engine room, go forward to the gun, get L'Heureux, and lead him back!" Then remembering that I had still a third man on the bottom, Tom Eadie, who had gone down after L'Heureux, I seized Eadie's telephone, asked him:

"Say, Tom, did you see anything of L'Heureux when

you went down?"

Promptly came a reply from Eadie:

"Yeh! When I hit the sub, there was L'Heureux standing by the gun, not moving, an' holding up that light in the water like he was the Statue o' Liberty!" Eadie paused a moment to catch up on his breathing, then continued: "Knowing it was L'Heureux's first dip, I clapped him on the back, laid my own helmet against his, an' asked him was he all right. I thought L'Heureux answered 'Yeh' so I left him an' went forward on my own job!"

So the pressure had apparently frozen L'Heureux into immobility or he had completely lost his sense of direction in the dim depths, and was afraid to move. Still there was the possibility that he had started, but had wandered forward instead of aft. I sang out into Eadie's phone again:

"Tom, go aft from where you're working till you find L'Heureux!"

Anxiously I waited, cursing inwardly. I now had Eiben coming forward, Eadie going aft, along the submarine, both wasting their dives. But I instantly forgot all that when Eadie reported:

"On deck! He's not by the gun! I just bumped into Joe here and he says L'Heureux isn't anywhere aft either!"

Startled, I pondered that incredulously. My two divers had met amidships, one from forward, one from aft, and no L'Heureux between them. While it seemed impossible on the narrow deck of that submarine, still they must have missed him. But on the other hand, both

were exceptionally good men, and that was highly unlikely. What then had happened to L'Heureux? With
a sinking feeling in the stomach, I tried once more to
raise L'Heureux on his own telephone. No answer.

And then from the depths came another message, excitedly ringing in my ears from Eadie's phone:

"On deck! I think I see a light out on the bottom,
about a hundred feet off the sub's starboard beam, and
it's going farther away all the time!"

"That must be L'Heureux, Tom! He must have
fallen off the sub and he's lost! Slide down the side
and chase him!"

Eiben lowered Eadie down the submarine's side, hand
over hand paying out his lines till Eadie struck bottom.
Then Eadie started out across the firm sands of the
ocean floor, breasting his way through the water following that will-o'-the-wisp of a glimmer vaguely seen
through the translucent depths. Finally he caught up
with it. There was L'Heureux, aimlessly wandering
over the sands, the brilliant light clasped in his hand
still flashing hither and yon seeking for the submarine,
while curious fish swam all about him, darting suddenly
in and out of the searchlight's gleam.

Eadie took his shipmate by the hand and guided him
back to the submarine, where Eiben hauled both of
them up on deck. Practically all of their allotted hour
on the bottom was gone; we signaled all three men to
stand by to rise. Nevertheless, on the topside I heaved
a deep sigh of relief, thankful that nothing worse than
two completely wasted dives had resulted.

We started up the three men for decompression. Diving was over for the day; we had no more divers to send down. As the late afternoon waned, step by step we heaved the divers up through the cold sea, carefully decompressing them, giving even a little extra time at each stop for L'Heureux's benefit, to insure thoroughly decompressing him. Finally, as darkness fell, with over two hours having elapsed since their rise started, we hoisted the three dripping divers in over the *Falcon's* side and dropped them on deck. The tenders rushed in to strip off their burdensome rigs.

I loitered near L'Heureux, curious to note whether he looked frightened after his queer mishap. Off came his helmet. I asked him how he felt.

"Fine, Mr. Ellsberg! Fine!"

No question about it; he wasn't bluffing. L'Heureux was positively bubbling over with mirth, as happy as if he had just come from a very live party. Unquestionably he had on what we called an "oxygen jag," a state of mild intoxication, similar to the alcoholic variety, induced occasionally in some divers by the exhilaration of the excessive oxygen breathed in heavily compressed air. It meant nothing and would soon pass off, but it did explain his queer silence and his otherwise inexcusable falling off the sub. The moment he had let go of the gun he had apparently staggered drunkenly overboard and as drunkenly had zigzagged away across the ocean.

Surgeon Flotte, our medical officer, and an outstanding authority on the physiology of diving, tall and lean, even muffled in his overcoat, came up, observed L'Heureux a moment, then confirmed my surmise.

"Oxygen intoxication," he muttered.

There was no gain in questioning L'Heureux. The cause of his mishap was evident. And there was even less profit in laying him out for having ruined our day's work, expensive as that loss was—it simply wasn't his fault if excessive oxygen had jazzed him up. Not to make him feel badly by questioning him further, I left.

The tenders finished undressing the divers, leaving the *Falcon's* fantail a mess of wet lead shoes, lead belts, helmets, hoses, and sodden diving rigs, and the three, still in their underwear, hurried below for supper, already somewhat late.

Five minutes later, seated on a bench before the heavily laden mess table, L'Heureux, still as merry as ever, suddenly collapsed, pitched forward on the table, unconscious!

No need to ask questions in that company. "The bends" and a bad case of it! No one wasted time in futile first aid measures. Hastily his shipmates seized the silent figure of L'Heureux, unceremoniously rushed him up the steep ladder to the deck above, while a cry rang out forward:

"Dr. Flotte! Dr. Flotte!"

In through the round steel door of the recompression tank went L'Heureux's inert form, one diver dragging his head, another pushing on his legs. Running from the wardroom came Surgeon Flotte, to dive through the opening almost on L'Heureux's heels. The door clanged shut behind him. On went the compressed air, hissing gently into the chamber as was customary. Hastily Surgeon Flotte felt L'Heureux. No sign of heartbeat. The man was completely out, might die at any moment, might perhaps already be dead from bubbles of air clogging his heart. It was no time for routine measures. At any cost those bubbles must be reduced to a size small enough to pass through the heart valves, to allow circulation to continue. And only high air pressure could compress them enough for that.

Dr. Flotte sprang for the air valve and twisted it wide open. Immediately the low hiss of the incoming air changed to a loud roar and, under the terrific pressure of the high pressure air banks, air started to pour into that recompression chamber. The needle on the caisson gauge jumped like a race horse getting away from the barrier, continued rapidly round the dial. Twenty pounds—40—50. Dr. Flotte's ears began to ring. That was as high in pressure as we had ever gone before on anybody. But no stop now. Sixty pounds. Blood oozed in Flotte's nose and mouth, but still he kept the roaring in full blast. He must get the pressure up on L'Heureux, never mind himself. Seventy pounds, with the valve wide open, the needle still racing up the dial. Eighty pounds, a higher pressure by far than anybody on that diving job had ever before been subjected to, either on the bottom or in the tank, and, worst of all for Flotte, taken in one swift rush!

Eighty pounds was enough. Flotte shut off the air. Dizzy from the sudden impact of high pressure, ears ringing excruciatingly, he bent over L'Heureux, tore off his shirt. The diver's chest was covered with purple splotches, the result of the bursting of a myriad small blood vessels from expanding air. But that was a minor

result of "the bends." The major question was circulation. Had he got those heart bubbles down before
L'Heureux's heart had stopped forever?

Flotte bent over his chest, listened, then smiled wanly.
His heroic treatment had succeeded. A faint heartbeat
became perceptible, L'Heureux began to breathe again.
The bubbles, compressed to one-sixth their previous
size by the sudden application of pressure, were passing
out of the heart; blood was beginning to pump through
it once more.

Gradually then Flotte began to release the air from
the chamber, decompressing L'Heureux by regular
stages. But in spite of working over him all night
through, in spite of everything that his medical skill
could suggest, Flotte was never able to bring L'Heureux
back to consciousness. Through the long hours he lay
there as the air pressure went down, limp, unconscious,
apparently paralyzed in some degree, simply breathing
feebly.

At 3 A.M., Dr. Flotte emerged from the recompression chamber, weak and dazed from his own exertions
and the shock of high pressure. He sought out Lieutenant Hartley, skipper of the *Falcon*.

"Everything that pressure can do for L'Heureux's
been done. Everything that I can do for him here has
been done. He's paralyzed and he's nearly gone. If we're
going to save L'Heureux's life, we've got to get him to
a hospital right away!"

The *Falcon* was 25 miles at sea from Newport and
the nearest hospital, swinging to her six hawsers mooring us over the grave of the *S-51*. Getting out in a small

boat to unshackle those six hawsers from the mooring buoys often took an hour. Hartley called me.

"We're getting underway for Newport at once with L'Heureux. No time to unmoor now. I'm going to cut the hawsers. O.K.?"

I nodded.

In a moment, from bow to stern, axes were flashing downward into the eight-inch manila hawsers that held us to our buoys, and the frayed ends of the hacked-off cables went flying overboard, the severed mooring lines drifting off in the waves like undulating sea serpents. Below, our forced draft fans commenced to shriek, pouring air into the furnaces as the oil burners, suddenly opened wide, began to spray huge quantities of fuel into the fires.

Quivering from keel to superstructure from racing engine and twirling propeller, we got underway in the darkness and headed under full power for Newport, radioing to the Naval Hospital there to have an ambulance waiting on the dock when we arrived.

At 7 A.M. in the early dawn, we transferred the still unconscious L'Heureux to the ambulance and sadly headed back to sea to pick up our severed hawsers.

That was mid-November. When we landed him, L'Heureux had been a man of something over 160 pounds weight. Within a few weeks, partial paralysis, including his kidneys, resulting from "the bends," had wasted him away to a skeleton of 70 pounds, and there for months he hovered precariously between life and death. Not until late the following July, after an eight-month struggle in the hospital, did he finally recover sufficiently to be discharged.

What happened to L'Heureux? Why didn't the de-compression we gave him, when he rose from that un-lucky dive, free him of nitrogen in the normal man-ner, save him from a disastrous case of "the bends"? We never knew; Dr. Flotte could never explain it. The man had once made a deeper dive than that and had come up unaffected. What was different on this occa-sion? We could only theorize. Something, a sort of mental shell shock so to speak, at finding himself on that smashed submarine, still the coffin of most of her crew, may have so upset his normal nervous and phys-ical reactions that his blood circulation may have been much reduced during his rise, with the result that he failed to decompress properly and came up with abnor-mal quantities of nitrogen still in his blood, which very promptly, once he was no longer under pressure, came out of solution and knocked him cold.

That, at any rate, is all we know. Haldane's decom-pression tables are correct, and we as divers follow them to avoid the dreaded scourge of diving. But once in a while in the abnormal case they fail to work, and "the bends" are with us again, as frightening as ever.

Abnormalities in diving are occasionally on the good side, leaving scientific theory just as unable to explain clearly why. Some few divers have worked under con-ditions which should each day have killed them with "the bends," and been no more affected than if they had never left the surface.

The most famous case was that of Alexander Lambert, a British diver, trained in the days when "the bends" was simply a mysterious demonic visitation and to whom, during most of his diving life, the term "decompression" would have been simply an unmeaning phrase. On his most noted exploit, the salvage of $350,000 worth of gold coin from the Spanish vessel *Alphonso XII*, sunk in 1885 off the Grand Canary, he had to work inside the vessel at a depth of 162 feet, 27 fathoms down. This was the maximum depth up to that time to which any diver had been called on to dive.

Lambert, the greatest diver of his day, and perhaps of any day, was a physical marvel. As described by one who knew him well and who often saw him stripped, he was a "medium-sized Hercules," a man only 5 feet 8 inches high, but with a chest measurement of 52 inches and with arms, legs, and a neck to correspond, giving him enormous physical strength.

In the course of his salvage work on the *Alphonso XII*, Lambert, single handed, blasted his way down through three iron decks to the treasure room, located just above the shaft tunnel, and then sent up the gold in boxes weighing some 200 pounds each.

Lambert made 33 dives on the job, usually from 2 to 3 dives a day. He averaged about 20 to 25 minutes on the bottom on each dive, coming up each time his work was done in from 4 to 5 minutes, just as fast as his tenders could heave his massive form up through the 27 fathoms of sea covering the wreck.

According to the usual decompression tables, to avoid "the bends" he should have spent about 60 minutes on

From "Deep Diving"; Courtesy Sir Robert H. Davis

ALEXANDER LAMBERT IN THE TREASURE ROOM OF THE
"ALPHONSO XII"

his rise, instead of only 4 or 5. And he should never have attempted a second, let alone a third, dive on the same day.

Why didn't "the bends" strike Lambert down after his first dive, as it would have done with any ordinary diver? And if it didn't get him the first time, how did he ever escape on his second and third dives of the same day, when the average man, even properly decompressed on each rise, would certainly have been stricken?

There is no certain explanation. In diving, that "blood is thicker than water" is more than a trite expression. Only upon the fact that blood *is* thicker than water rests our stage decompression theory. Since blood, unlike the charged water used in the ginger ale experiment before cited, is a thick, viscous fluid, bubbles do not form in it until the pressure has been more than cut in half. This fact makes it possible to hoist the diver quickly from whatever depth he is working at to a point where the pressure is about halved, where he is held while some nitrogen escapes from solution, without dangerous bubbles forming.

Now Alec Lambert was an extraordinary human being, possibly a decidedly abnormal one. It may be that his blood was much thicker than the average man's, thus still further slowing up the rate at which bubbles would form in his veins and arteries. Aside from that, his enormous chest gave him a lung capacity and a circulatory system far above ordinary, which may have enabled him to eliminate nitrogen from his body at such a rapid rate that a few minutes spent in rising rid him of nitrogen better than an hour's decompression served other divers. On top of all, it seems probable that with his gigantic body he had a heart to correspond, a heart so large that an amount of entrapped air, due to lack of decompression, that would have killed an ordinary man, never noticeably incommoded him.

Whatever the reasons, this modern Hercules, Lambert, made a record for immunity on 32 deep dives never since equaled. His regular daily performances without decompression were such as would make a modern diving supervisor stand aghast should one of his men suggest repeating even one of them.

But, in the end, even Lambert pressed his luck too far, as is only too apt to happen in the presence of sunken gold. With $350,000 in gold coin salvaged from the *Alphonso XII* and $100,000 remaining in the water-logged chests left in the wreck, Lambert imprudently lengthened his thirty-third dive to 45 minutes, instead of rising after his usual 25. For a 45-minute stay on the bottom, our present tables call for a decompression time of about 90 minutes. Lambert, as always, rose in less than 5 minutes, but this time even his magnificent body was unequal to the strain. "The bends" got him at last. Half an hour after rising, he became suddenly and completely paralyzed from the waist down, and from that paralysis, to the end of his days, he never completely recovered.

CHAPTER IX

AsIDE from caisson disease, a deep sea diver faces other troubles. One, briefly mentioned before, is "oxygen intoxication." Under a pressure of 5 atmospheres, for instance, corresponding to a depth of about 22 fathoms, with each breath he takes a diver gets five times the habitual quantity of oxygen to which nature has accustomed us. This unusual dose of oxygen, carried to the brain and the nerve centers, gives them extraordinary stimulation and in some individuals, unduly susceptible, produces an effect indistinguishable from alcoholic intoxication.

L'Heureux exhibited this effect, but he was not unique in that. Another diver on the same salvage job, and a competent worker previously in shallower water, exhibited such marked symptoms of an "oxygen jag" on his first dive in deep water that his mate, a veteran deep diver, after watching him staggering drunkenly over the heeled-down deck of the *S-51*, had to hang on to him to keep him from reeling overboard. Finally, despairing of getting any work done while clinging to an inebriated helper, he signaled us on the surface to haul them both up. That man we never used again except as a tender on deck.

A third case was somewhat different. This man, a diver with long experience, never failed to do a good job on the bottom, but once he hit the submarine, he invariably developed an "oxygen jag," singing to himself as merrily and as drunkenly as if he had one foot hooked over a brass rail hoisting in Scotch, and conversing, when he had to, over the telephone with every symptom of being "half seas over." When he rose, he was still that way—vivacious, bubbling over, and absolutely happy, until the effect of the oxygen wore off, when he relapsed into his normal morose state.

For oxygen intoxication there seems to be no remedy other than to eliminate from the diving roll those unduly susceptible and liable to be useless on the bottom or to endanger themselves.

Excess oxygen has one other effect, even on those who are not intoxicated by it. Under a 5-atmosphere pressure, the diver gets as much oxygen per breath as if on the surface he was breathing in an oxygen tent, inhaling only pure oxygen. Breathing pure oxygen unduly stimulates the subject. For that reason it is given in some critical hospital cases to save a patient. The result is much the same as seen when an object is burned in pure oxygen rather than in ordinary air (which is diluted with about 80% nitrogen). Combustion is far more rapid, the flame is far more brilliant, and the heat developed is far more intense.

Effectively then a diver under heavy pressure is acted on similarly to the object being burned in pure oxygen. The combustion of his tissues is accelerated, the fat in his body is literally burned out of him after a few deep dives, and generally, thus stimulated, he is able to perform feats of muscular strength under pressure on the bottom that would be beyond him on the surface. This is sometimes unfortunate when a diver is tempted to try some superhuman tasks, for it gives his body an over-strain which normally he could never subject it to.

The only diver ever lost in work with me died as a result of this. James Frazer, chief torpedoman—6 feet tall, 200 pounds in weight, blond, blue-eyed, placid; in appearance, temperament, and intelligence, a modern Viking and at the time one of the Navy's veteran divers —came to the S-51 wreck first of all our salvage crew to reach there. Throughout the early months on that job, Frazer was the main reliance for work both in-side and outside the submarine. Working with him was Francis Smith, another chief torpedoman who in ap-pearance was almost Frazer's direct antithesis—slight in figure, dark, sparkling in temperament.

These two opposites made a perfect diving pair, part-ners in their every dive almost from the beginning; and the dangers these two went through in working on that smashed and corpse-filled submarine make one of the all-time epics of diving. But by courage and cool-headed thinking in moments of desperate peril both Frazer and Smith came out of every visible trap unscathed. What finally got Frazer was an intangible peril he never sensed.

We had to seal up an open hatch on the S-51 to make her watertight. For that purpose, I had made a sizable round steel plate about an inch and a half thick, fitted with a large valve on top and a heavy rubber hose be-neath, long enough to reach the bottom inside the sub-marine's central compartment. Through the hose we would expel the water inside her when the hatch cover was in place and the compartment sealed off.

The completed hatch assembly weighed about 500 pounds, far too much for even two men to handle. So to get it into place, an improvised derrick and a half-ton chain block were rigged out from the conning tower of the submarine to plumb the hatch opening before Frazer and Smith finally dived to install the hatch assembly in the opening.

From the *Falcon's* main boom, we lowered the heavy hatch cover into the sea, guiding it down to its destination by a line which had been attached to the conning tower, but in the open ocean the rolling of the *Falcon* was so great as to prevent our dropping the load where we wanted it on the *S-51*. At last, after several failures to land it properly, the disgusted Frazer signaled us from the bottom to let go, which we did. But the heavy mass landed finally not even on deck but on the high rounded side of the submarine alongside the superstructure where it perched precariously, too far away from the conning tower to hook it with the derrick chain fall and hoist it into place.

In that predicament, Frazer got out on the side of the submarine and, assisted as much as possible by Smith tugging on the rubber hose, almost single-handed he heaved that 500-pound cover several feet up over the superstructure to the deck, a really Herculean feat which on the surface he could never have performed. Oxygen stimulation alone lent him the strength, but many times since that hour I have fervently wished that Frazer had never seen either that submarine or that hatch.

Shortly after being hauled up from the depths and aboard the *Falcon*, Frazer complained of a pain in his chest. Surgeon Flotte examined him, then with a grave face informed me that Frazer had badly dilated his heart, that he must never dive in deep water again. We put him ashore in the hospital. After treatment there he came out some weeks later. The next few years he was in and out of hospitals under treatment, and then his heart collapsed, and at thirty-six, James Frazer, as fine a diver and as fine a shipmate as one might ever hope to meet, suddenly died, a victim of what an unseen peril, oxygen stimulation, had done to him.

A few other hazards come from pressure, the worst of which is the "squeeze." To balance off the sea pressure, as explained before, the diving rig must be kept supplied with air at a pressure slightly greater than that of the surrounding sea. But to avoid getting too much buoyancy, only the helmet and a little space in the suit just below the breastplate actually contain any air, the remainder of the rig from there down being folded tightly in against the diver's legs and body by the water round about.

The pressure of the air in the lungs, however, is instantaneously communicated through the blood to every part of the diver's body, in a sense inflating him to counterbalance the weight of the sea, and thus he normally works. If, however, the pressure in his helmet is decreased, even slightly, a peculiar sensation results. Many times while learning to dive, in balancing off my rig I have felt it. Let only a little too much air be bled off from the helmet, so that the internal body pressure is lower than that of the water, and the sea starts to press it in from every direction, giving one the weird sensation of being tightly hugged all over, a sea embrace so perfect, so all-inclusive of every inch of body that it might well be the envy of every siren who ever lived. If a little more air is released, the hug of the sea becomes tighter and tighter, forcing the blood out of the legs first, then out of the lower abdomen, giving an "incipient squeeze" which is both painful and hazardous. Even so far, no diver ever voluntarily dares go.

But a real "squeeze" never comes more than once to any man. In that case, usually as a consequence of a fall under water to much greater depth, the pressure suddenly increases on the diver as he goes deeper, before he can get enough additional air into his helmet to balance it. The result is instant and horrible death. The sea pressure, considerably exceeding the pressure in the helmet and the lungs, forces the blood from limbs and body in one vast hemorrhage into the chest cavity, and then, taking the diver's body encased in his flexible dress, it rams him up into his inflexible copper helmet as if his frame had suddenly been put under a hydraulic press, molding him into a jelly that completely fills the helmet and leaving the rest of his body an unrecognizable mass of mangled flesh and protruding bones.

One instance of this occurred some years ago in nearby coastal waters. An ocean liner, sunk there in collision, was being examined by divers, one of whom foolishly walked far forward on her sunken hull so that his lifelines no longer tended up and down to the salvage ship but ran in a long sweeping curve through the water down to him. Standing on the bow of the wreck, somehow he lost his balance, whether from the swift tidal current, whether from a misstep, was never known. Overboard he went and, with his lifelines slack and nothing to support him, plunged immediately to the bottom. The increased water pressure promptly "squeezed" him, and what the tenders finally hauled up inside his suit could not ever have been thought once a man.

A second case was more peculiar, in that it occurred not even in deep water, but under the shadow of New York's skyscrapers where no one might have expected a fatal diving accident. While we on the *Falcon* were working out at sea in 22 fathoms, momentarily expecting death to strike at some diver struggling in a shattered wreck, a small commercial job was going forward in the East River, where, in only 5 fathoms of water, two divers were working on a submerged power house bulkhead. They were diving from a small deckhouse scow from which a ladder led down to their work, with a pump inside the deckhouse to provide the air which, at no great pressure for that slight depth, they needed. On the scow one man only casually tended both divers, whose air hoses were coupled up to a common manifold from the pump.

Ashore the noon whistles in Brooklyn began to blow. Time to eat. The tender went over to the diving hoses, yanked each one as a signal to the diver below to come up for dinner, and then, leaving his two divers to climb up the ladder from the muddy river bottom to the scow, he busied himself elsewhere about its deck.

A few minutes later he came back to the diving station. A trail of mud and water now led from the top of the ladder into the deckhouse. Concluding that both divers were up, he uncoupled their diving hoses from the pump manifold, and then entered the deckhouse to help undress them. To his horror, inside that deckhouse there was only one diver!

Hastily the tender leaped outside, coupled up again the second air hose, frenziedly signaled to his man below. But there was no response. That brief interval with the hose uncoupled had been more than enough. The diver's air had all blown out of his helmet through the opened hose into the atmosphere above, leaving him without any pressure whatever, and the sea, slight though the depth was, had instantly pounced on him, "squeezing" him to death. A mangled corpse was all that came up when the anguished tender heaved in on the lifeline.

To minimize the dangers of a "squeeze" the careful tender never for any reason lets go the lifelines of his diver from the moment he goes overboard till the time he is safely back on deck. We trained our tenders to "fish" their divers' lines, that is, continuously to test the slack of the lines by taking them up a trifle, as a fisherman sometimes does when he thinks he has a bite. If the lifelines showed slack, it was the tender's job then to take them in till he could just "feel" his diver. It is not possible to keep the lines taut, for then the diver would find himself unable to move about, nor is it always possible to keep the lines up and down, for the diver working below may go some distance from the point of his descent, but too much of a sweep to the lines of a man working under conditions where a fall means a sharp drop into deeper water should always be avoided.

It is best before sending a man down under such conditions to move the diving ship to plumb approximately the required working point, and never to allow the sweep of the lines to exceed a slight amount, so that a diver falling will bring up dangling at the end of his lines before he has dropped far enough to permit the sea to give him more than an "incipient squeeze." Of course on exploratory dives of new-found wrecks, this isn't always possible till after the wreck has been well marked out, and the initial divers must take their chances as, with their lifelines trailing them on a long sweep, they go over the wreck.

While not so serious, next to "the bends" nothing ever troubles divers so much as their eardrums and certain head cavities. Air pressure here is, as usual, the villain.

Our eardrums are exposed to the air through external cavities, as every person very well knows who has ever gone swimming and got water in his ears. What is not common knowledge is that the eardrums on their inner sides are also exposed to the air, through intricate internal passages called the Eustachian tubes, leading from the nasal cavity to the drums.

As a diver goes down and the air pressure builds up in his helmet, the increasing pressure immediately smites his eardrums from the outside. If all is well with the diver, the pressure inside his nose works with only a slight lag through the Eustachian tubes to the inner side of his drums, leaving him with nothing more than the trifling feeling of ear strain that we all experience in a rapidly descending elevator or a train suddenly entering a tunnel.

But if all is not well with the diver, his ears promptly let him know in no uncertain tones, with a ringing and sharp pains that warn the veteran diver instantly to halt his descent. What is happening is that his Eustachian tubes are wholly or partly clogged (usually from a cold), and the air pressure is not equalizing on both sides of his eardrums. This condition causes those unfortunate membranes to stretch and bulge inward under the excess pressure outside them, just as the rubber in a toy balloon bulges when it is blown up.

The effect is excruciatingly painful. Nothing can be done except to stop and clear the Eustachian tubes so that the pressure may equalize and the drums snap back to normal. In ordinary circumstances this is not difficult, since all that is necessary is to hold the nose closed and then to swallow sharply. (Try doing this a few times and from the way your eardrums snap in and out you'll quickly learn how your own Eustachian tubes are operating.) But the diver unfortunately cannot quite do this, for his nose is cased inside a copper helmet where he can't get his fingers on it to close it off. All he can do is to press it against the faceplate of his helmet as firmly as possible and then hope for the best while he swallows hard.

If he manages to clear the passages enough to relieve his ears he can continue down, repeating the process each few feet of his descent to keep them so. But if he fails to clear them, he goes deeper at his peril, for his drums will only bulge inward further as more unbalanced pressure comes on them, and then rupture completely, leaving him deaf. The wise diver, if he fails to clear his ears, comes up and ceases diving till the cause of his trouble is healed. The foolhardy diver disregards his instruction, disregards the warning pain, and keeps on dropping. I had one such once, a supposed veteran, who, on his first dive, intent on showing his new shipmates how good he was by an extraordinarily rapid descent to the bottom, managed to burst both his eardrums and put himself completely out of the job.

And then finally there are the skull cavities to cause trouble, those cavities around and above the nasal passages (and communicating with them) which are known as the sinuses. These cavities, intended by nature in some way to help moisten the inspired air and influence the resonance of the voice and perhaps for other purposes of which I am yet unaware, are linked with the nostrils by narrow passages. I never knew that such things as sinus cavities or passages existed till one of my own unforgettable early diving experiences vividly and literally impressed the knowledge on my mind.

It was winter time. I was working in a naval diving tank some 16 feet deep, experimenting with an underwater torch to cut steel, which I hoped to develop in time to use on the salvage of the S-51 the following spring when we could resume diving at sea.

The tank being shallow, we dispensed with some of the accessories used in deep sea work. For instance, we omitted the telephone set, as the diver was visible to his tender, only 16 feet above him, and furthermore could be constantly watched through glass ports near the bottom of the tank. Moreover, the depth being only a few fathoms, no decompression at all was needed, and it was possible, instead of telephoning, to come up a dozen times if necessary to say what was wanted or to make changes in the torch without loss of valuable hours in rising.

I turned up one January morning with a nose running from a cold, but, not wishing to waste a diving day, dressed to go down in the tank. Just before my helmet went on, I held my nose closed, gulped vigorously several times to get my Eustachian tubes well opened, and then, with the helmet hastily screwed on, dropped hurriedly down to the bottom of the tank before those ear passages could clog up again with mucus and cause me trouble in descending.

For some 30 minutes, I worked on the bottom of the tank under what might be called ideal diving conditions, clear fresh water, comfortably warmed, and no possibility of danger. Above me, Bill Carr, bosun's mate first class, sat on the top of the tank, smoking his stubby pipe the while he tended my lines and through the clear water watched with deep interest my none too successful efforts to get the submerged torch to make some impression on the steel plate I was endeavoring to burn in half.

Concluding finally that I should have to make a further change in the torch design, I shut off the torch and then, having inflated my rig a bit to make the lifting easier, I jerked three times on my lifeline—the signal that I wanted to be hauled up. Above me, Bill Carr laid aside his pipe, braced himself on the platform topping the tall tank, seized my lines firmly with his two powerful hands, and started to heave in. My lead-soled shoes immediately lost contact with the tank bottom and up through the water I went.

I had not risen a yard before a severe pain, the like of which I had never before experienced, darted across my forehead just above both eyes and promptly became unbearable. Whatever it was I didn't know, but unquestionably decreasing water pressure while I rose was causing it. Frenziedly I shouted out:

"Stop heaving, Carr!" But there was no telephone cable connected to my helmet, and on the topside of that tank, Bill Carr, completely ignorant of my plight and unable to hear my shouts, placidly shifted his grip and heaved me up another yard.

Remembering suddenly that I had no telephone, I quit calling and endeavored to signal. Clawing upward, I vainly tried to jerk my lifeline twice, the signal to slack off, but with my whole weight dangling now on that line it was so taut the jerks failed to telegraph upward to Carr's calloused fingers and he continued to heave in.

Never have I felt such pain. Just above my eyes it seemed as if the top of my skull was about to blow off, and the farther I rose the worse the torture became. Thank God, I had only 16 feet of water to rise through in helpless agony before my helmet popped through the surface, and the lift ceased. Dazed and speechless, when my helmet was removed and I could get my hands on my head, my arms shot involuntarily upward and grabbed the top of my cranium, pressing it down to relieve the torment, while Carr, flabbergasted at my strange action, stared at me in dumb amazement.

No more diving for me that day, nor any day following till my cold was gone. When the pain had subsided enough so I could bear it, Carr finished undressing me and I moved on to consult our surgeon. From him I learned the mysteries of sinuses and discovered what had happened to me. Through my sinus passages, clogged with mucus from my cold, the compressed air had gradually worked its way into the sinus cavities during my stay on the bottom till the pressure in them had reached a balance with the air I was breathing. That, of course, I never noticed. But when I started up, the same clogged passages prevented the compressed air from escaping easily, while the pressure around me decreased and left me with my sinus cavities full of expanding compressed air trying to do exactly what it felt like—blow the top of my head off in an effort to escape. Since that day, I have always felt a brotherly sympathy for the fellowship of sinus sufferers, of which clan I have discovered there are unfortunately a goodly

number, and with them all I can fervently say that sinus
pains without dispute top the list of human agonies.

A few more queer things result from diving pressures. A fairly obvious fact is that as air is compressed
it gets heavier—at 2-atmospheres pressure weighing
twice what it normally does; at 3 atmospheres, three
times as much; and so on proportionately. This has a
peculiar effect on speech. We form our sounds by the
action of the vocal cords, the teeth, the tongue, the
palate, the lips, and the nose, on the air escaping by
them. But our human organs of speech are designed for
and accustomed to working on air of ordinary weight,
and strange things happen to our speech when the air
starts to get unduly heavy. Not the least of the human
organs which are no doubt astonished at what their
owner demands of them in deep sea diving are the
vocal cords when they find themselves trying to
vibrate harmoniously in a "thick" medium for which
they were never designed.

When the pressure gets somewhat over 3 atmospheres, at depths beyond 90 feet, the diver's vocal
troubles start. He is unable to form his sounds normally with his speech organs working on such heavy
air, and his words, as heard on the surface over a telephone, often sound "mushed up" and sometimes unintelligible. As an example, so heavy is the air below 90
feet in depth that a diver cannot audibly whisper (as-
suming that he ever wished to do so unlikely a thing)
and he cannot whistle. In neither case can the lips vi-

brate the dense air enough to make the slightest sound.

A noticeable effect in deep sea work is that the voice loses all its individual quality. It becomes difficult to distinguish one man's voice from another's, and it often takes an excellently trained and imaginative telephone tender to figure out what his diver is trying to say. I have been on both ends of this dilemma. As salvage officer on the surface I shouldn't wonder that I have nearly driven some divers crazy by my cross-examining them over the telephone, making them repeat in different forms the information they were trying to convey, to make sure I understood what they were saying. And as a diver myself on the bottom, after having slowly, painfully, and carefully articulated my words to make a message conveying my wishes clear to my telephone tender, I have often wondered how any human being could be so thick-headed as the dumbbell at the other end of my line, and I have sometimes been driven to voluble profanity in my efforts to make myself understood over a diving telephone.

There is no remedy. Speech is badly mangled under heavy pressure; only an experienced diver can begin to make himself understood, and only a highly experienced telephone tender—who knows what the diver is trying to do on that particular dive, who knows the conditions of the wreck about him, and who has a flexible imagination capable of guessing at what the diver is probably talking about—can piece together from the few intelligent sounds, in the jumble coming up from the bottom, what it is that the diver is saying. But still the average diver will solemnly assure you that the stupidest sailor on earth is the man on the topside end of his telephone line.

Last but not least of the dangers a diver faces is that of "blowing up." Naturally enough, like any other body placed in water, a diver must weigh more than the water he displaces or he will float. While the average human body is slightly heavier than water and will of itself usually sink (an unfortunate trait as most of us have discovered in our efforts to learn to swim), once a copper helmet displacing a considerable volume of water is clapped on, a sizable amount of ballast in the form of lead-weighted shoes and a lead-weighted belt must also be draped on the diver to make him heavy enough when submerged so that the slightest current in the water does not waft him away like a feather in a breeze.

It is up to the diver on reaching bottom so to regulate the amount of air inflating his flexible suit as to keep him heavy enough for the work at hand but not so heavy that dragging his weighted form over the ocean floor unduly wears him out. But whatever he does, every diver is trained never to lighten himself so much that he becomes positively buoyant and commences to float. If he does, the dangers he faces are extreme.

The man who by inexperience or miscalculation sets his inlet and exhaust air valves to keep a little too much air in his suit immediately begins to float upward through the water, but such unfortunately is the weird world in which he is working, that he does not always immediately note the fact. As he rises a little, the water pressure on him decreases somewhat, due to the decreased depth, and the air in his rig expands still further, inflating his suit a little more and of course increasing his buoyancy and accelerating his rise.

Up to this point, a quick-thinking diver can still save himself by instantly and completely shutting off his incoming air and opening the exhaust valve in his helmet wide by bumping it with his chin, thus releasing enough air to deflate his rig and allow him to sink again; but woe to the man who doesn't recognize his danger in the first fathom of his rise and promptly act to correct matters. For after that, he is helpless. The expanding air quickly swells out his suit, so inflating it that its ordinarily flexible canvas and rubber fabric becomes as rigid as sheet steel, and he finds himself utterly unable to bend his distended canvas sleeves enough to get his fingers in on his breastplate and shut off his air valve.

As he rises a little further and the pressure inside his suit still more exceeds the water pressure, his rig balloons out completely and becomes as stiff as a board. Looking now like a submarine blimp, he starts to shoot for the surface under the impulse of his enormous buoyancy. What happens after this is in the hands of Fate.

Should he come up beneath the diving ship, he will ram it, going like a torpedo, and dash out his brains. If, by luck, he misses the hull of the ship above, he will burst through the surface, shoot up into the air a moment like a leaping tarpon and (provided nothing has happened to his suit) fall back into the waves to be reeled in by his tender like a hooked fish.

If he has been on the bottom long enough to soak

up much nitrogen before "blowing up," his rocketlike rise without decompression will insure him a case of "the bends," unless his mates can promptly put him under pressure again. But if his canvas suit bursts under the internal pressure, either on his way up or when he breaks surface, his situation then is most desperate. Immediately the air in his rig escapes through the rip, he promptly deflates like a punctured balloon, and, with all his weights dragging him down, sinks like a rock, his lifelines naturally being badly slacked from his rise and unable to stop him before he hits bottom again. And that finishes him. With no pressure in his rig to save him as he goes down, the sea promptly "squeezes" him up into his helmet or else drowns him.

I have seen three cases of "blowing up," one occurring with a novice and two with divers who, if they were not entitled to the name of veterans, then there are none such in this world.

The first mischance happened to a young seaman who trained as a deep sea diver in the same group with me and under the same instructors. Later, after I had myself made several dives in deep water on a wreck, it came this bluejacket's turn. His task on his first dive was simple, merely to pick up off the deck of a sunken submarine a small copper pipe elbow which a previous diver had carelessly set down on the sub's deck as he was preparing to rise and had then forgotten to bring up with him. We badly needed that copper elbow for reworking in our pipe shop before the next job inside the flooded submarine could proceed.

The tenders dressed the youngster, a boy of about eighteen or twenty, and just before the helmet was dropped over his tousled head, I gave him his instructions. Within six feet of where the descending line would land him on that submarine's deck, he would find this copper elbow lying alongside the engine room hatch. He was to pick it up, walk back to the descending line, and come up again immediately. His entire time on the bottom was to be not over a minute.

The diver nodded his head. He understood. That his youthful face looked a little pale as the helmet came down and locked him inside the rig was excusable enough, for no matter how much training a new man gets, there is always something terrifying at the prospect of the first actual encounter with the deep sea.

Overboard went the diver, and down the descending line as his tender paid out his air hose. Straight down with no delay he dropped into the depths, and length after length the lifelines went out after him over the rail. The hose slacked in the tender's hands, stopped running. The tender started to "fish" the lines.

"On the bottom!" sang out the man tending the telephone line.

I waited a little nervously, for any man's first dive is always a nerve-racking ordeal for the salvage officer on deck as well as for the diver himself. A minute went by, and then a second one, and still no signal from the diver to start him up.

I turned to the telephone man.

"Ask him, 'What's the matter?'"

The telephone tender leaned over his transmitter, repeated the question, listened a moment, then looked up at me puzzled.

"I can't make out what he says, sir."

That wasn't surprising. I gave up bothering with the telephone and stepped over to the tender handling the lifelines.

"Give him 'one,' " I ordered. One jerk on the life-line means "Are you all right?" and a return jerk of "one" from the diver indicates "I am all right."

The tender at the rail obediently jerked the lifeline once, then, much surprised, replied:

"I can't 'feel' him, sir! The lines seem too slack!" He immediately began to take in slack so he could signal.

Slack lines? That was odd. No extra slack had been paid out, since the diver needed hardly to move from where he landed on the submarine. But before anything more could be done, the telephone man excitedly burst out:

"I can't make out what he's sayin', sir, but he just let out an awful yell!"

On top of that, the tender at the rail sang out:

"Something's wrong on the bottom, sir! I've got in a couple o' fathoms o' hose already, and the lines are still slack!"

Something was wrong, all right. One continuous shriek was echoing in the telephone; at the rail, the tender, frantically heaving in lifeline, was unable to catch up with the slack. Then on the surface to starboard of the *Falcon* long coils of black air hose started to emerge, tangling in snaky loops among the waves. And immediately following them, shooting vertically upward into the air like a skyrocket, flashed into view our diver!

With suit ballooned out and arms spread-eagled

stiffly sideways, before our startled eyes he hung poised a moment above the waves like a seaplane taking off, water dripping copiously from his grotesque figure. With a loud smack he then fell back into the sea, floating horizontally high in the water some 60 or 70 feet off our rail, and rising lightly to each passing wave.

"Heave in! Heave in!" I roared, fearful that his overstrained suit might burst any instant, leaving him with all that slack hose there in the water between him and us, to go plummeting back into the depths.

No urging was necessary. As fast as human hands and arms could work, the hose came in, to be piled, coils, snarls, and all, in a heap on deck. And at the end of the hose, towing helplessly face downward in the water, came our diver. Over the side at the end of the boom went our stage with two seamen on it, to be dropped instantly waist deep into the waves. They floated the prostrate diver in over the stage between them, and, still prone, we hoisted him aboard, shut off his air for him, yanked him upright, and deflated his rig. When we removed his helmet, he was absolutely white and so petrified with fear as to be speechless. As a precaution, we rushed him into the recompression chamber and gave him a moderate decompression in that, though there never was any chance of "the bends" getting him, for he had not been on the bottom long enough before "blowing up" to soak up any noticeable quantity of nitrogen.

Except in being cured of any desire to dive again on that job, he showed no ill effects from his startling rise and next day was back again on his former job as dresser to the other divers. I believe he appreciated his luck in missing the *Falcon* coming up. He was completely unable to explain how he came to set his air valves wrongly or anything else that went on on the bottom, remembering only his horror when his suit bulged suddenly out and he saw the submarine disappearing from beneath his feet.

A second case of "blowing up" came on the *S-4* and happened, even more inexcusably, to a man who had been diving over fifteen years, being (with several others) at that time the holder of the world's record for depth in deep sea diving. He, a veteran of many diving operations, on his first dive on the *S-4*, practically duplicated the experience above, first improperly adjusting his air, and then within a minute of reaching the *S-4*, finding himself helplessly spread-eagled in his suit and rocketing surfaceward.

That both these men missed killing themselves by coming up beneath the ship and ramming her at high speed was not wholly a matter of luck, for we always made it a practice to hold the *Falcon*, if possible, a little clear of the wreck below against just such contingencies. But the third case I observed was not so happy in that respect.

We were engaged in driving a tunnel through the clay and sand beneath the *S-51*, a difficult, long-drawn-out, and nerve-racking business for the divers engaged in it. For weeks on end, one after another our divers stretched themselves out flat in the hard clay under that submarine in the confined tunnel, in absolute blackness and with a mixture of icy water, mud, and sand swirling around them, boring their way through by washing out the clay ahead with the stream from a fire-hose nozzle which it was their job to direct. A job more trying on a man's nerves it is difficult to imagine and only some six of our most experienced divers ever made much progress at it. An hour in that tunnel, with the smashed submarine overhead and the dangers of under-water tunneling ominously present all around, was more than any ordinary diver could stand.

Tom Eadie, one of our experts, had on this occasion completed his hour's work and was on his way to the surface decompressing. With him standing on the lit-tle steel stage, some two square yards in area, was Joe Eiben who, having by chance finished some other task at the time Eadie emerged from the tunnel, was being brought up with him. The stage was at the 90-foot depth, the first stop in decompressing a man after he had left the bottom.

There suspended on a line dangling from the end of the *Falcon's* boom and vertically beneath our gunwale, 45 feet to the bottom and twice as far from the sur-face, with nothing but the gray translucence of the water visible in any direction, up, down, or sideways, the two divers, experienced men both, began their de-compression. As our routine demanded, both com-menced exercising vigorously to stimulate their circula-tions and thereby accelerate the elimination of nitrogen from their systems. On deck, we began to get the next diver dressed and ready to slide down the descending line to pick up Eadie's discarded hose and resume the tunneling. Then came a call on deck from the tender

Courtesy Commander Henry Hartley, U.S.N.

DIVING STAGE WITH STEEL SUSPENSION BAILS USED ON
U.S.S. "FALCON" TO DECOMPRESS DIVERS ON THEIR WAY UP
TO THE SURFACE. DIVER JOE EIBEN GOING OVERBOARD

on Eadie's telephone:

"Tom's just sung out to me but I can't make out what he wants!"

That was queer. With the men at 90 feet, they should just be coming into the range where the thinning air should no longer make speech difficult. I took Eadie's telephone, but while he certainly was saying something, it was completely unintelligible to me. I passed the telephone set to Lieutenant Hartley, then to one of our gunners, but neither of them was any more able than I to make head or tail of the jumble of sounds in the receiver. Something seemed wrong with that telephone.

An idea struck me. Joe Eiben was on the tiny stage alongside Eadie. I took Eiben's telephone set and called out:

"Hello, Joe! Ask Tom what he wants!" That would quickly settle it, for all Eiben had to do was to lay his helmet against Eadie's, get the answer, and phone it up to me. I listened intently. And then came a shock.

"Tom's not here!" replied Eiben. "What did you pull him up for?"

Pull him up? Nobody had pulled him up to my knowledge. I glanced sharply at the sailor tending Eadie's lifelines.

"Where's Tom?" I asked, wondering if by some freak he had lifted his diver up off the stage.

"Still down there, sir. I'm trying to signal him. I've given him 'one' on his line two or three times already, but he don't answer!"

What was wrong? Eadie vanishing literally from Eiben's elbow without the latter's knowledge, shouting unintelligibly over his telephone, and failing to answer signals! It all added up to trouble, worse because his inexplicable disappearance from that platform was so fantastically unbelievable! A sudden leaden feeling griped my stomach. What was happening to Eadie?

And then in my ears came a shout from Eiben far below me in the sea:

"Eadie just fell back on the stage! His suit's nearly torn in half and he's full of water! Take him up quick!"

Eadie's rig torn open and Eadie drowning 15 fathoms down? No time for decompression now. Action was swift. Half a dozen sailors seized Eadie's lines and began to heave in, but the weight on the end of those life-lines was tremendous. Evidently his suit was completely waterlogged, with no buoyancy left. More men grabbed the lines wherever they could get a handhold and we heaved in desperately lest Eadie drown before we got him up. Meanwhile another stage was hurriedly dropped over the side with several men on it and lowered partly into the water. And then in breathless suspense we waited while hand over hand Eadie's lines came in over the gunwale.

A copper helmet at last rose into view. The men on the stage seized it and dragged it and the limp form inside the torn suit attached to it onto the steel plat-form. Up shot the stage, inboard, and then down with a clatter on deck, while all the while water gushed out of the canvas rig draping the prostrate figure on the platform.

The diving rig was a mess. With the canvas suit torn nearly in two just below the breastplate, with the leather

straps which held the helmet down to the belt both
parted, and with the heavy lead belt draped round the
ankles of the rig, nothing was in place except the shoes.
There was no need to unscrew the helmet and unbolt
the breastplate. We simply cut loose the shoes and
dragged Eadie out of his rig through the gaping hole
in its breast.

Eadie was very pale, half-strangled, bleeding badly
from the mouth and nose as well as from a gash on his
chin, but apparently still semi-conscious. We waited
for no investigation there on deck; this diver yanked
up from the depths after a long dive needed all the
decompression we could give him. The tenders who
pulled him out of his rig hustled him, without a pause,
into the recompression tank, and we ran up 50 pounds
pressure of air on him, while Surgeon Flotte began to
dress his cuts and check his other injuries.

What had happened? I learned hours later, when
Eiben, who finished his decompression on the stage,
had reached the deck and joined Eadie who then lay
wrapped in blankets in his bunk below.

On the little stage at 90 feet the two divers were
exercising in the water to help decompression. Eadie,
with his rig somewhat lightened up to lessen the weight,
was jumping up and down on the stage which hung
from triangular steel bails secured to lines several
fathoms above the platform itself. Eiben was doing
knee stoops. Each man, interested solely in making sure
he was ridding himself of nitrogen, paid little attention
to the other.

Suddenly the automatic exhaust valve on Eadie's

helmet jammed—unfortunately for him, in the closed position—and the inevitable happened. His suit, with the air escape now shut off, immediately started to swell out, his buoyancy increased, and up through the water he started. Eadie, a very quick-thinking diver, recognizing the danger, tried at once to reach the control valve on his breastplate and shut off his incoming air; but so fast were things happening that, before he could swing his gloved hand in on his chest to shut that valve, his suit had stiffened out so from internal pressure that it spread-eagled him and, with arms helplessly stretched sidewise, for all the world like a farmer's scarecrow, he started to "blow up."

Eadie, in real peril now, still kept his head. Unable to do anything for himself, he shouted into his telephone transmitter:

"Turn off my air on deck!"

That his telephone tender, receiving suddenly from the depths such an utterly unheard-of order from a diver, failed completely to understand him, is natural enough, and things with Eadie were happening too fast for him to repeat his request. He was completely off the stage and rising rapidly.

At that point in Eadie's narrative, I looked from Eadie, still pale and wan and with his jaw swathed in bandages as he lay swallowed up in the middle tier of a nest of sailors' bunks, to Eiben, who was seated on a stool alongside me in the narrow passage.

"Say, Joe," I asked, "what was the matter with you? You were so close to Tom it's a wonder you two didn't knock each other off that platform exercising. Couldn't

you see him 'blowing up' and grab his feet, at least,
to stop him as he went by you?"

Eiben, by nature that unusual phenomenon among
sailors, a man who always strictly minded his own business, looked at me sheepishly. This time his very whole-
some trait had worked out wrong, and he well knew it.

"Yes, commander," he muttered, "I sort of saw him
go by me out of the corner of my faceplate, but I just
figured he was making an extra-high jump there in the
water, and I went right on exercising. I wasn't thinking
about Tom at all and I didn't look around again for
him till you called me from the deck."

"Queer," I thought, but everything is queer in the
deep sea and there was no cause to upbraid the unimagi-
native Joe for not visualizing an unimaginable situation.
Whoever could imagine a diver like Eadie "blowing
up" once he was safely on that platform?

I glanced inquiringly at Tom's worn face and he
continued weakly to explain.

Still shooting upward, he thought fast. He knew, of
course, that, after a long dive, if he rose suddenly without decompression, he might finish with "the bends,"
but at the moment that worried him little. What terri-
fied him was the knowledge that the stage he had just
left was suspended from the *Falcon's* side, and that he
might well come up beneath her, and, propelled by all
that buoyancy, going as if he'd been fired from a gun
by the time he hit her hull, with every prospect that
the impact would flatten his helmet like a pancake and
finish him then and there.

As he shot upward, completely helpless now, he saw

through his faceplate the top of the steel bail from which the stage hung, flashing downward past him. He could do nothing to seize it with his spread-eagled arms, but instantly he thrust toward it both his weighted feet, and, with inexpressible relief, managed to hook the brass toe caps of both his diving shoes into the little triangle at the peak of that bail! He stopped with a violent jerk that nearly tore his feet off his ankles, and there he hung by his toes, the tremendous buoyancy of his bulging rig straining to start him upward again the while he prayed fervently that the stubby toe caps on his shoes might not tear off!

Once more he tried to bend an arm in and shut off his air, but it was impossible. He thrust his head forward and bumped the inside of his exhaust valve with his chin, trying to free it and let the air escape, but it was jammed hard and he could not budge it. Meanwhile with more air entering all the time, hammered down to him by the compressors, the pressure in his already fully distended suit rapidly increased further, putting a fierce strain on the leather shoulder straps holding his helmet down to his belt. The heavy straps burst. No longer restrained, his canvas suit suddenly stretched upward, the helmet and breastplate rising instantly over half a yard. Firmly anchored down by his feet, he was unable to rise with it as his rig stretched out, and the copper collar of the breastplate as it leaped skyward past his neck hit him a clip on the chin that nearly broke his jaw and left him half-stunned and bleeding. Meanwhile, with nothing to support it any longer, his 80-pound lead belt dropped down round his feet, while the suddenly extended suit, more buoyant now than ever, increased sharply its lift, and·the pull on Eadie's toes became almost intolerable.

Eadie, dazed though he was by the blow on his chin and with his tortured body stretched as though gripped on a medieval rack, still kept his wits about him. Momentarily, at least, he had stopped himself; it did not seem possible that it could be for long. The telephone was now his only hope of aid. He threw back his head, shouted upward:

"On deck! Tell Joe to climb up to me, shut off my air and open my petcock!"

But with the telephone transmitter inside his helmet, and that helmet now two feet above him, it was not strange that no one on deck ever made him out.

The pressure inside Eadie's rig began to climb rapidly, nearly bursting his eardrums, and he commenced bleeding from both nose and mouth. The strain on his feet increased, he lost all hope, and was wondering only how much longer it would be before his toe caps tore loose and he went rocketing upward to his death—when abruptly his suit burst!

Instantly in one vast bubble, the air escaped to the sea, and his bulging canvas suit, completely deflated, shriveled in on him in the twinkling of an eye. Eadie nearly burst as the extra pressure suddenly left him, eardrums, sinuses, lungs almost exploding under the sharp reversal of tension. And then, no longer supported, down dropped his helmet, water poured into his suit through the gaping rent over his breast, and, with no further buoyancy to sustain him, his lead ballast promptly dragged him downward, releasing his toes from their grip beneath the steel bail. Like a rock, he dropped down through the water, fortunately landing in a waterlogged tangle of torn canvas, lead, and copper on the stage at the feet of the dumbfounded Eiben!

Joe Eiben undoubtedly saved Eadie's life. Had he not been there instantly to warn us of his shipmate's desperate predicament, Tom Eadie would unquestionably have drowned in his own diving rig before we got him up.

But even so, Eadie's perils were by no means then over. What his feelings were, 15 fathoms down in a ruptured and flooded suit, dazedly struggling to avoid swallowing any water, can well be imagined. When on top of that he sensed that he was being heaved upward by lines attached only to his breastplate, while he himself lay limply huddled in a flooded suit which was already torn half across just below that breastplate, with the lead belt tangled round his feet and his lead-weighted shoes dragging downward on the torn fabric, his mental agony was complete. If the canvas let go entirely, with no lines whatever attached to his severed suit, he would plunge to the bottom hopelessly ballasted down by those terrible weights about his feet. Spasmodically he shook his legs as he was jerked upward, trying to shake loose the belt and thus ease the strain on his torn rig. But he was unsuccessful, and, fearing finally to make matters worse by his struggles, he lay quiet while we heaved him madly toward the surface.

And so we got him up. Curious, after his recital, I examined his exhaust valve to see why it had jammed and, by causing him to "blow up," had started his nerve-racking and rapid-fire series of facing death in so many different forms in a space of probably not over a minute. On disassembly of that exhaust valve, the trouble was immediately obvious. While he was lying prone in the tunnel working in the mud with his fire-hose, some muddy water, washing into his helmet past one side of his valve while air blew out the other, had carried a little sand into the valve housing. This sand, when he was coming up, had finally lodged in the valve sleeve, tightly jamming the valve stem and preventing it from sliding open. Just a few fine grains of sand, that was all, requiring a microscope almost to see them well, but they had come within a hair's breadth of killing one of our best divers.

So it goes. Between "bends" and "squeezes" and the other real dangers arising from air pressure and water pressure, the diver always has enough to be concerned about without wasting time worrying about the legend-ary monsters of the deep which furbish the average diving tale.

CHAPTER X

PROBABLY of all paradoxes in human experience, not one seems crazier or more unbelievable to the average mind than the practice of lighting a torch at the bottom of the sea and then proceeding with that torch to burn through steel wreckage immersed in water. Yet that miracle is now as everyday a fact as radio, and probably somewhere at this moment a diver is prosaically working with the Ellsberg torch in slicing through thick steel cofferdams at the bottom of some arm of the ocean, removing the steel piling which, when intact, held back the water, while inside the cofferdams, far below the level of the sea outside, concrete and masonry foundations were built for another huge bridge.

The conception of flame under water may, for all I know, be quite as old as that other ancient conception of man, flying through the air, with which in the heroic days of Homer the Greeks entertained themselves. But the practice, like the practice of heavier than air machines outdoing the birds, had to wait for engineering science and modern ingenuity to bring it to reality.

That necessity is the mother of invention was amply demonstrated in the case of the underwater torch. Something workable simply *had* to be produced to insure success in salvaging the *S-51*, where I first came into forcible contact with both the problem and the

Courtesy Craftsweld Equipment Corp., N. Y.

DAMAGED SECTION OF 48-INCH DIAMETER CAST IRON WATER MAIN 2 INCHES IN THICKNESS CUT 50 FEET BELOW THE SURFACE WITH THE ELLSBERG UNDERWATER TORCH. THIS WATER MAIN,

need.

Wrecks are always a mess, and each one is different. However well-ordered a ship normally operating on the surface may be, let her be rammed in a collision or ruptured by an internal explosion, and then dropped heavily to the sea floor, while water, under increasing pressure, floods her compartments and collapses bulkheads which temporarily may resist, and the condition of the ship as she hits the bottom—badly heeled over on her side, perhaps capsized completely, or sometimes broken in two—is always a sickening sight to those who must work on her or in her. Decks slope so steeply that no man can walk on them; they may be even vertical. Corridors and passages are no longer normal; a man may have to crawl along the bulkheads on hands and knees because he can no longer stand erect in a passage lying on its side. Steel doors no longer open, or sometimes, when they must be closed, cannot be shut; because in either case they are so jammed with wreckage or sprung by twisted bulkheads that the hinges will not swing or the door will not seat properly on its frame. Wreckage is everywhere, barring progress. The slightest task for the diver often becomes an almost insuperable problem.

What to do? The diver and the salvage officer do the best they can. That they ever get results is a miracle. There is no successful wrecking job the real story of which is not a vivid romance. But often, in spite of all, the salvagers fail, and in those failures are human tragedies as heart-rending as any Shakespeare ever wrote.

On the bottom of the sea a man's capacity to do anything is small compared with what it is on the surface. Mechanically he is handicapped by a bulky and cumbersome diving rig which makes it difficult if not impossible to get into tight places or through small doors. He can hardly work over his head, for the breastplate makes it hard to lift an arm more than shoulder high. It is hard to see, for the light is usually dim; frequently inside a wreck it is completely black. Even the strongest submarine lamps usually help little, for, if the mud is stirred up outside a wreck or the water inside is oily and dirty, the lamp becomes worthless. I have often seen occasions when it was necessary, even with a 1,000-watt searchlight (a bulb so powerful that it could be lighted only under water lest otherwise the heat melt the glass), to thrust it smack up against a diver's faceplate in order to determine whether the light was burning or not. In such cases a diver can only work by feel, and if the water is cold and he must wear gloves to avoid freezing his hands, the sense of feel is no great help.

Worst of all, aside from mechanical difficulties, is the impossibility under heavy pressure of working with a clear mind, of thinking rationally, and sometimes even of keeping the thoughts on the work in hand rather than on the very evident surrounding dangers and the chances of avoiding them in getting safely back to the surface.

So it is evident that an obstacle in a diver's path on a wreck is magnified out of all proportion to what the same obstruction would mean to him were the ship normally afloat, and any means of removing or of cutting it away assumes a value to the diver far above what a similar tool is worth to him on the surface.

I learned all these things while attempting to salvage the submarine *S-51*. That was a vessel of some 1,200 tons, smashed open in a collision, lying badly heeled over to port in 22 fathoms of water about 14 miles out in the wild Atlantic. The salvage job looked fairly hopeless, but, because of peculiar circumstances surrounding the loss of the ship and her crew, the Navy was going to try. Many salvage experts both at home and abroad looked on the undertaking as completely futile, as simply a waste of time and money, intended only to still the criticisms leveled at the Navy Department, and never possible of success.

Partly by luck and partly by persistence, I managed to get myself designated as salvage officer. I was a naval officer, a naval constructor, and a naval engineer, quite confident that certain engineering apparatus I intended to bring to the task would soon lift that hulk to the surface. Nor was I wholly wrong. The engineering apparatus I used was quite adequate to make the lift, but I quickly saw that it was highly doubtful that it would do it *soon* or indeed *ever*, for getting that apparatus attached to the hulk was proving a Sisyphean task, and I concluded that I unfortunately had rushed in where angels fear to tread.

The difficulty was, of course, the divers. I had had no previous experience either with diving or divers, and, having checked the fact that men could work in 22 fathoms of water, had taken it for granted that what a mechanic could do on the surface, he could still do, fairly well, encased in a diving rig. Sad to relate for me, it turned out not to be so. I quickly learned that most divers then were very poor mechanics, and that it was not possible to put a good mechanic into a diving rig and have a good diver—at least not till after months and perhaps years of training him to feel somewhat at home in the water.

In this dilemma, then, there was nothing for it if the job was not to be a failure save to try to make better mechanics of the men who were good divers already. I fully realized that, even if they became experts, there were only a few of them available and that, even in the ideal case of the expert mechanic who was also an expert diver, his ability to work effectively on the bottom was less than that under ordinary conditions of a mediocre apprentice just learning to use his tools.

So in this wise, I struggled along some weeks, getting sickeningly little done on the bottom compared to what was necessary to attach the lifting gear and raise the submarine, though I quickly came to a deep respect for the divers for what they were accomplishing under terrible handicaps.

In the midst of this situation, we had a mishap. When we attempted to lower our first pontoon alongside the stern of the submarine, the lowering hawsers broke, dropping the pontoon (a weight of some 40 tons) about 80 feet to the bottom of the sea, where it landed happily clear of the submarine, but with the two lifting chains and wire hawsers attached to that pontoon and intended to go under the sub, a dismal tangle of snarled wire and chain about the S-51. Before we could proceed, it was necessary to get that pontoon to the surface again and get that mess of wire and chain clear of the sub's stern.

One massive chain (its links weighed about 100 pounds each) and wire hawser we managed to drag clear by having a diver with a sledge hammer drive out the 2½-inch thick shackle pin which held chain and wire hawser together. But with the other tangle we could do nothing, for so great had been the strain on the cable there, that the thick steel pin had bent and would not drive free of its shackle and release the second wire.

No way remained except to cut something. As between cutting in half either a 2½-inch thick anchor cable link or a 1-inch thick steel wire, both buried in the sand on the ocean floor, prudence suggested trying to cut the wire, even though the anchor link was made of soft wrought iron while the wire was of hardened plow steel.

We had with us what then amounted practically to a laboratory curiosity, the Navy design for an underwater torch more or less similar to several others which had also been tried in Europe. In addition, we had as its operator a young diver from the Brooklyn Navy Yard, George Anderson, who had worked with its designers in the Naval Laboratory and of all people knew most about its use. Here was a job just made for that torch. All George Anderson had to do was to go down, light off the torch, burn that wire in two, and our troubles were over.

We dressed Anderson, and, to help in holding the

wire if necessary while Anderson burned it, we also dressed another diver, John Kelley, a chief torpedoman and a naval diver of long experience. They were a queerly assorted pair, young George Anderson looking even more youthful than he was with his smooth complexion and his thick, curly hair, while John Kelley, almost bald, his face tanned and wrinkled from long exposure to the sea and his teeth darkened from much tobacco, looked old enough to be George's grandfather.

Then having rigged up the torch and the igniter and tested out both diving telephones, we clapped the helmets down on the men and one by one lifted them over the side. They slid down the descending line to cut that wire.

Plodding aft on the ocean floor alongside the partly buried submarine, they came to the starboard propeller, its bronze blades still gleaming dully through the water. Here they turned at right angles and followed the trace of the wire hawser till, some 40 feet away, with the submarine no longer visible, they picked up the loom of the pontoon, showing vaguely through the water like a vast cylindrical tank standing vertically on one end on the bottom of the sea. There in the dark shadow of that pontoon, half buried in the sand, the wire hawser joined the heavy anchor cable which hung down from a hawsepipe in the pontoon with the ponderous links of the chain, originally intended for battleship anchor cable, piled erratically in a heap on the sand.

The wire, unlike the chain, had no loose coils lying around. It ran straight from the pile of cable to the keel of the invisible submarine, being completely hidden in the sand for most of its length. Kelley bent over, dug his fingers into the sand to get a grip round the wire and heaved up hard to get some slack to give Anderson something to work on, but the wire proved to be stretched taut and he was unable to get an inch. That proving unavailing, Kelley dropped to his knees, and with his gloved hands dug a hole in the sand beneath the wire, exposing it fully close beside the chain link to which it was shackled. Rising again, he pressed his helmet against Anderson's, pointed to his little excavation, and shouted through the water to his mate that the job was ready.

The slow bobbing of Anderson's helmet in the water indicated that he understood. Somewhat awkwardly he knelt in the sand, signaled for a little more slack on his lifelines so he could stay that way, and readjusted his air valves to give less buoyancy so that he would not continuously have to fight too much of a lift to his rig in keeping his head down.

Kelley meanwhile, standing erect beside him, telephoned up to us:

"Anderson's ready! Turn on the torch!"

On deck of the *Falcon*, we opened the valves to the various banks of gas bottles and closed the switch on the electric igniter circuit. Below, Kelley pressed the contacts of the igniter together over the tip of the torch and released the contacts. After several futile attempts, he finally got a spark to flash, and the torch lighted. On the surface, listening over Kelley's telephone, I could hear it banging away in the water.

On deck, I waited and listened while Lieutenant

Hartley watched the gas bottles. To us waiting uneasily on the topside, the minutes fled rapidly by, and it seemed the wire should long ago have been severed, but we got no report of completion. Bottle after bottle of gas emptied itself out down the hoses to the torch and fresh ones were connected up, and still no report. I was disgusted. Wire should be much simpler to burn than steel plate or iron bars, for each individual strand of wire, being small, heats up almost immediately to the burning point.

Our gas supply was nearly gone, two thirds of an hour had gone by and there was no signal yet that the wire was cut. I was anxious for news, but, not wanting to interrupt Anderson who obviously was busy, I took again Kelley's telephone set and slipped the receivers over my ears.

The chattering of Kelley's teeth came to me clearly over the wire. Kelley was evidently very cold, standing there motionlessly in the icy water close alongside Anderson, not daring to move or to exercise lest he upset his partner, and fearing to shift farther away because Anderson might need him. Kelley was plainly enough freezing in silence.

"Hello, John!" I sang out into his transmitter. "How's Anderson making out with the torch?"

Immediately Kelley shut off his air to make himself heard better, then answered:

"Mr. Ellsberg, if I could only take my helmet off and get my teeth on that wire, I could chew it in half faster than that damned torch is burning it!"

A loud roar echoed in my receivers as Kelley turned

on his air again, abruptly ending the conversation. I laughed sourly, recalling the state of Kelley's teeth, but Lieutenant Hartley, when I repeated the reply, only looked the more glumly at the gas gauges. His last bottle was emptying fast. In another minute it would be completely gone.

Then came a welcome call from the bottom.

"On deck! The last strand's cut! Take us up!"

But that didn't quite end the matter. Kelley, too chilled to decompress properly, developed a bad case of "the bends" and we had to keep him under treatment all night long in the recompression chamber to relieve him.

That ended any dreams of expediting the job by slicing wreckage with the torch. A man with a hacksaw could have cut that wire much faster, not to mention that the amount of bottled gases used was enormous and quite out of question for practical results. We gave up the torch and struggled along.

But when, in December, bitter cold and continuous Atlantic gales drove us off the job till spring, I decided that if only I could invent a torch that would work my chances of successful completion would be measurably improved, and I turned to on the task. A major reason for previous failures to produce such an instrument seemed to me to be self-evident—that the laboratory engineers who formerly had worked on the problem were not divers, and no divers were engineers enough to work it out.

I was not a diver either, but I was an engineer, and if only I became a diver, I might not only solve the torch problem but be able also actually to supervise and direct the rest of the salvage work on that submarine from the bottom of the sea, where I could myself size up each task rather than have to take second hand my knowledge of requirements and possibilities through the eyes of men never trained as engineers. I had little choice anyway, unless I were willing to see that job end in dismal failure, as a previous attempt to salvage a sister submarine had a few years before. So I became a diver.

John Kelley, Bill Carr, and Jim Frazer (the latter temporarily out of the hospital with his heart feeling better) were my instructors, and a tougher lot of teachers a man never had in anything. But they knew their stuff practically, and what they knew they taught me. What they didn't know about diving (which was mostly technical theory), I picked up from Lieutenant Hartley, Surgeon Flotte, and the earlier experimenters in the field of pressure effects.

Working in a narrow 16-foot-deep tank in the Navy Yard, I soon learned enough of diving to start my torch experiments, and all through the winter I worked at the bottom of that tank with innumerable torch designs, developing the basis for a practical one.

We all know from common observation that many materials heated up in air take fire and burn; some of us know that those same materials plus some others that cannot be ignited at all in air will burn much more fiercely if immersed in pure oxygen undiluted with the nitrogen so plentifully present in air. Moreover, it is a commonplace of everyday life that, if you want to put out a fire, the simplest way is to douse it plentifully with water and the next simplest perhaps is to shut off the air supply.

How then does it happen that the bottom of the sea, the very first place one would choose to put out a fire (where there are literally oceans of water around and no air at all to feed the flames) is indeed a practicable place not only to start a fire but to go much further and burn through steel which is ordinarily incombustible?

The theory is simplicity itself. Provide (artificially of course) everything necessary for combustion, and keep the water away from your blaze, and you can have your fire, no matter how much water there may be in the near vicinity. It is just as simple as the problem of combining sufficient engine power and correctly designed and controlled wings to permit heavier than air flight.

For combustion under water it is only necessary to bring together two gases that will burn, hydrogen and oxygen, for instance, and keep the surrounding water away from the flame so that it will not cool it enough to extinguish it, which is what usually happens when we pour a bucket of water on a fire and put it out. Mechanically this aim is achieved by piping down to the torch through one hose a stream of hydrogen gas and through another, a stream of oxygen, mixing these two gases in proper proportions in the torch head, and then lighting off somehow the mixture, which now requires nothing whatever additional for combustion. To keep away the water, a stream of compressed air piped down through a third hose is ejected in a hollow cylindrical sleeve all about the torch tip, so that no water ever comes near enough to the flame to cool it unduly.

The above is all theory. Getting a flame under water is not so difficult. Creating a torch that would deliver a flame that would set steel on fire under water and slice it through and thereby be of some practical use was quite something else, as George Anderson's experience very well illustrated.

I struggled with torch designs and operation and with various gases as substitutes for hydrogen, which (in no slang sense at all) is not so "hot" when it comes to combustion. About all I got out of it for weeks was headaches, a bad cold, and the top of my skull nearly taken off, as I have previously related. There were many obstacles. Hydrogen doesn't give nearly as hot a flame in combustion as does acetylene, which is the gas ordinarily used for torch work on the surface. With hydrogen, a flame temperature of 4,000° F. is usual, while acetylene gives a flame temperature of over 6,000° F.

Acetylene, however, suffers from the unfortunate possibility of breaking down, when stored under much pressure, into compounds that are likely to explode spontaneously. As any gas to be used at the bottom of the deep sea must be pushed down to the diver under considerable pressure or it will never reach him, acetylene, in spite of its attractive heat, had to be ruled out unless I wanted to take a chance of killing the divers with torch explosions. Since the divers faced dangers enough already without adding this one, I had, though regretfully, to drop acetylene.

I tried many other strange gases as substitutes for the unusable acetylene, in an effort to develop really hot flames, but all suffered from one drawback or another, and ultimately I was driven back to relying solely on hydrogen as a fuel. So I accepted that as inevitable and the problem came down to getting a torch which would develop a flame hot enough to cut with hydrogen, oxygen, and compressed air.

Much of the work I did myself on the bottom of that tank, but not all. Often Kelley or Frazer manned the torch, while, through the glass ports outside, I watched carefully to see how each torch design worked, directing the diver over the telephone as to what adjustments I wanted him to make. (By this time, after my sad experience with no telephone when I learned about sinuses, we had gone back to telephone rigs again.) Endeavoring to overcome the relatively low heat of burning hydrogen, I tried out varying torch designs to enlarge the size of the flame; and, working with a huge torch one day, the biggest I have ever seen anywhere, we came near disaster.

Frazer was in the tank. His deep sea diving days were definitely over on account of his heart, but he objected strenuously to giving up diving altogether, and, as 16 feet was practically no depth at all, and there was no physical exertion involved in the experiments, the doctor saw no objections to his working there, while I saw plenty of gain, for I have never run across a more competent or more intelligent diver than big Jim Frazer.

Through a small glass port in the side of the tank,

giving a sort of aquarium view, I watched while Frazer worked inside. Before him in the water were a couple of wooden sawhorses of the kind ordinarily used by carpenters, but heavily weighted down with lead to keep them from floating up. Spread across the top of the horses was a steel plate, perhaps an inch in thickness, on which he was to try the new torch. The water inside the tank, while clear, was quite dark, for very little light came down from above, and the glass ports below were so small as to pass only a negligible quantity through them. Still that made no difference, for the operator would have light enough as soon as he ignited his torch.

And so it was. The moment Frazer, cased in his diving rig, flashed an electric spark from the igniter across the tip of that torch, it lighted off with a terrific bang and commenced burning with a beautiful blue conical flame, much larger than any we had had before. It brightly illuminated the steel plate beneath it and Frazer's grotesquely helmeted figure above it, casting a gigantic black shadow through the water behind him against the inside walls of the constricted tank.

I looked through the port, glowing with elation. With a flame as big as that, we should get heat enough to produce startling results; and so we immediately did, but not in the way I had anticipated.

As Frazer bent forward to bring the torch down on the steel plate, there was a slight puff, and the flame at the tip vanished as if sucked up inside the torch, leaving the interior of the tank as vaguely lighted as

CHIEF TORPEDOMAN JOHN KELLEY BEING DRESSED FOR A DIVE,
WITH CHIEF TORPEDOMAN JAMES FRAZER (REAR)

before. Through the dark water, I could see Frazer's body straightening slightly as he lifted the torch close before his helmet, his surprised eyes staring at it through his faceplate evidently unable to make out what had happened. And then, convulsively, Frazer's fingers suddenly opened, as if he had just discovered he was holding a rattlesnake, and he flung the torch from him violently, hurling it as far away as he could—which was, however, not very far, as the tank was only about 4 feet across one way and not over 8 feet the other.

Nonplussed at this strange behavior, I was wondering what childish fit of temper, so unlike the man, had abruptly gripped Frazer just because his torch had gone out, to make him throw it angrily away. Then came explanation.

Almost instantaneously, a radiant ball of fire, indescribably dazzling, at least 2 feet in diameter, blazed out there in the water inside the tank, flaming with all the fierce brilliance of the noonday sun! Immediately, as if under illumination of a thousand searchlights, the water lighted up, outlining Frazer in his diving rig spasmodically throwing his huge figure against the far side of the confined tank as if struggling to break through the steel plates and get away. It was hopeless, of course, but with the very flames of hell licking toward him, threatening, immersed though he was in water, to incinerate him, who could blame him for trying to smash through that steel side?

Half blinded by the tremendous light and horror-stricken by Frazer's danger, I staggered back from the port, ran sideways a few yards till I could see the tender on the platform atop the tank, shouted wildly:

"Pull up the torch, Bill! Yank it up!" Then I ran back again to the port.

Obediently Bill Carr started to heave up on the torch hoses and I watched that terrifying mass of flame rise through the tank past Frazer's huddled figure till, thank God, it cleared his helmet, leaving him again surrounded only by water, after which I scrambled madly up the ladder to the tank top to lend Bill Carr a hand with the fire.

My bosun's mate, when I got up there, was hesitating over what to do. Each time he dragged the burning torch near the surface, the flames billowed out over the tank top and threatened to burn him; when he lowered it a little, the fire, instead of being quenched by the water, only caused the water to boil furiously so that it looked as if Frazer in the tank below might well soon be cooked. In this dilemma, Kelley, who was still on the ground below, helped matters by running over to the gas bottles and shutting off the valves. This promptly reduced the size of the blaze enough so that Carr, with one wild heave, pulled the flaming torch out of the tank and sent it sailing like a fiery comet far out into the air, to land well clear of everything, where, when the gases in the hoses were exhausted, the blaze sputtered out at last. And then, hand over hand, we pulled Frazer up through water already far too hot for comfort, to find him deathly white and dripping perspiration as if he had been in a Turkish bath.

What happened? So far as I could learn from our

quivering diver, he had thought the torch had simply
gone out on him and was looking at it to see why, when
suddenly it seemed to him he had hold of a red-hot
iron. Instinctively he dropped the torch, and a moment
later saw it burst into a Vesuvius erupting flame and
scorching steam at his feet. As nearly as I could ever
determine, the flame at the tip must have flashed back
through the unusually large gas ports into the handle
of the torch, there to burn invisibly a moment till it
made the metal handle too hot to hold, as Frazer's
burned fingers mutely testified. Then going back a little farther, the flame had burned through the rubber
hoses carrying the gases, thus loosing under high pressure the two streams of hydrogen and oxygen with
startling results.

The condition of that huge torch bore out this theory. When it had cooled enough to be examined, we
found it a ruin, with its brass interior a mass of melted
metal, while only the compressed air hose remained attached to its handle, the two gas hoses being completely
burned away.

That ended all experiments with such monster
models. I concluded that I had better stick to smaller
torches and less likelihood of flashbacks, while I strove
to get a hot enough flame some other way than through
mere size. Meanwhile I could only be grimly thankful the shock of that accident had not proved too much
for Frazer's already weakened heart.

We lost a day while we emptied the diving tank of
its steaming contents and refilled it with colder water

from the city mains, after which we turned to again with one of my previous smaller designs.

The weeks dragged along, the winter wore away, and we obtained some real results. I worked out a design that seemed satisfactory and taught my divers to adjust for a certain blueness of the flame which indicated maximum combustion temperature. With such a combination we got some amazing results in slicing through inch-thick steel plate. Kelley did well with it, I did well with it, but burning steel under water did not seem to be included among Bill Carr's natural aptitudes and he never got anywhere, while, perversely enough, Jim Frazer, who would never be able to take a torch down into deep water, far outclassed everybody. Big Jim, in no wise feazed by his earlier experience, nonchalantly kept his huge body huddled over the roaring torch, slicing through steel as if it had been butter. On one occasion he cut 13 feet of heavy steel plate in 10 minutes—roughly about 1 foot every 45 seconds—a really amazing performance to watch, and, it need hardly be added, a very gratifying one to me.

Early spring came, the winter weather moderated somewhat and I decided to give up further diving in the tank and practice only under more nearly actual service conditions, by diving from one of the docks in the sea water of the Navy Yard itself. So we moved our gear from the enormous structural steel shop which housed the diving tank to the caisson gate of the yard's largest drydock, and there in the deepest water which the yard afforded, around 35 feet, we resumed our practice.

One of the Navy Yard derricks lowered down for us a shot of heavy anchor cable of the size we intended to use for the lifting cradle of the S-51—battleship chain with the links formed of wrought iron bars 2½ inches in diameter, links so ponderous that a man wanted a crowbar to move even one of them.

When all was ready, John Kelley and I were dressed and lowered down into the water to cut that chain, while on the caisson gate an interested crowd of spectators gathered to witness the performance. I am afraid, however, they were disappointed in the show we put on, as they never saw anything after we dropped into the sea except air bubbles bursting on the roiled surface, for the East River water was vile—dirty, soiled with sewage, and so thick with sediment that even a few inches below the surface a diver's helmet faded into invisibility.

To make matters worse, when we reached the bottom, Kelley and I found ourselves sinking to our lead-belted waists into gooey mud, which our slightest movements soon stirred up into clouds of black ooze that made the water thereafter as impenetrable to sight as if a giant squid had emitted a vast stream of sepia ink directly enveloping us.

The water there, so different from our comfortable tank, was frigid as well as muddy. The cold I rather welcomed, as being a good introduction to what would be normal diving conditions out in the Atlantic. The mud, however, I could thankfully have dispensed with, as, from the diver's reports, we had a fairly clean sand bottom over hard clay where our submarine lay, but there was no help for it. With our diving lamp completely useless to us in the murky water, Kelley and I, keeping actual touch with each other lest if we parted we should waste our entire dive endeavoring to regain contact, started to grope in the slime and the darkness for the chain, two mud-plastered monsters clinging to each other with one hand the while with the other each fumbled blindly in the ooze—a queer situation, I thought, for any naval officer whom the public ordinarily visualizes as one of a tribe of rather supercilious beings glittering in gold lace against a background of gleaming brightwork and holystoned teak on a flag-decorated quarterdeck.

Kelley found the chain, dragged me over to it, put my hand on it, then, holding the lamp against it, tried to illuminate a link about opposite our chests which hung in the water just above the muck in which we were half buried. The lamp was useless. By thrusting my helmet against it, I could just make out a vague glow in the utter blackness around the searchlight reflector that showed the lamp was actually turned on, but I couldn't see the chain link.

Disregarding the lamp, I felt out the link with my numbed fingers to get an idea of how it hung, then with electric igniter and torch held just before my faceplate so that I might properly adjust the torch flame to the proper blue color as usual, I lighted off.

Immediately I got an unpleasant surprise. The torch lighted off, all right, but the flame there, an inch away from my faceplate and not two inches from my startled eyes, was not any shade of blue, proper or improper, but something I had never seen before in all my tank experiments—it was a very vivid orange! And no matter what I did to it in adjusting the proportions of hydrogen and oxygen fed that flame, it never changed the slightest, remaining always the same unvarying color, a beautiful orange glow.

I felt a little sick. On the varying shades of blue as a guide to the one proportion of hydrogen and oxygen that gave a flame hot enough to set the steel on fire, I had absolutely depended and trained my men and myself to rely. Now that blue color and all its variations were completely absent. Had I been able to lift one of my lead-weighted shoes through the mud, I would have kicked myself for an utter fool. For long ago as a student I had learned, and should have remembered, that what I had been depending on was impossible in the sea.

What I was looking at was simply a repetition of an elementary chemical experiment—the coloring of any flame with sodium. The sea was full of salt—sodium chloride—and there before me, beautifully colored by the salt of the sea, was a lovely orange-tinted sodium flame! I had so far been misled by carrying out all my experiments in a tank filled with *fresh* water, a variation from actual conditions which had never occurred to me before as having any significance. But it was obvious now that it was going to have plenty, for the open sea where my submarine lay was far saltier even than the brackish tidal waters of the East River.

With my mind about as numbed by this shock as my fingers were by the cold water, I adjusted the flame by guess and felt for the chain link. Carefully I swung the torch tip against the chain, shoved my helmet forward till that orange flame was crackling almost in contact with the glass faceplate so that I might observe the action, and tried to heat the metal to the ignition point. I failed, and no adjustment I could make, now that my color guide was gone, got any results. Finally, I passed the burning torch to Kelley and pushed him over toward the chain for him to try, but his luck was no better then than mine. After a disheartening hour in the mud, the darkness, and the cold, we turned out the torch and signaled to be hauled up. I ran my fingers over the link before the tenders started heaving me upward. So far as touch could tell, we had not left so much as a dimple on the surface of that iron link to show for all the flame we had played against its smooth exterior.

I came up first. From that depth, no decompression was necessary, and I was hauled up smartly to clamber wearily up the last few feet on a ladder secured to a float alongside the caisson gate, where the tenders unscrewed my helmet. Blue with cold, I slumped to a seat on the edge of the float, my mud-plastered rig dripping with slime while I dangled my feet in the water to clean them somewhat before the tenders went to work to undress me further. Frazer bent over me inquiringly.

"How'd it go, commander?"

Dejected, I shook my head, too worn and too utterly dispirited to answer otherwise. The work of months seemed to have gone all for naught; the tool on which I relied most heavily for future success was a failure in sea water.

Then Kelley was hauled up. Holding with both hands to the ladder at the side of the float, he paused a moment, half in and half out of the water, while, in his turn, the tenders seized his helmet, with a sharp twist, broke the watertight joint to the breastplate, and, unscrewing his bulky copper hood, lifted it off him, leaving his bald head sticking grotesquely up through his breastplate while he blinked his eyes, accustoming them once more to the light of day.

From a few feet above us, clustered on top of the drydock caisson, the crowd of spectators leaning over the railing looked down curiously. Among them, brought down for the first time at her eager insistence to observe what her father was doing, was my three-year-old daughter, clinging tightly to her mother's hand and gazing wide-eyed, absorbed in silent contemplation of the queer clothes I was wearing and of my mud-plastered condition. But when my diving partner's helmet came off, her baby eyes promptly focused on the only matter of importance and she gurgled delightedly:

"O-o-h, look, daddy! Mr. Kelley's worse than you! He hasn't got any hair at all on the top of his head!"

Kelley's shining pate, a second before blue with cold, promptly turned a fiery red, and in the laughter of the crowd, the tragic effect of sodium on my torch took wings from my mind as I watched the blush spread from Kelley's face to the roots of what was left of the hair round his ears. Perhaps life wasn't so serious after all, and I had to laugh also as Kelley's face, sticking up above his copper collar, became a turkey red. Then, somewhat embarrassed myself, I twisted awkwardly about to indicate somehow to little Mary Phillips Ellsberg that polite children do not comment in public on such personal matters; but I was too late. An even more embarrassed mother had already removed her from the railing and was hurrying her across the caisson homeward bound for the other side of the Navy Yard.

Back to our diving tank we all went next day, and, with the tank filled with *sea* water this time, I started from scratch again. My problem now was to provide mechanically the controls that the diver needed on the torch, so that from touch and feel of them alone, without regard to the color of the flame (and indeed even without seeing it), he could get the optimum heat out of an oxy-hydrogen mixture.

In April, we sailed once more on the *Falcon* for the spot in the Atlantic where our submarine lay. Work on her—once we had buoyed her off again for our moorings and tied to her new descending lines forward, amidships, and aft—started with a rush. And very soon after came the first opportunity to try out the torch on a job, which, if successful, meant a radical reduction in diving labor—the task of securing the pontoons to the cradle of lifting chains threaded under the submarine.

The first dive I ever made in the deep sea was for that purpose. Frazer, our best torch operator, while still with us as a diving supervisor, couldn't dive in deep water any more. Kelley, certainly as expert as I with it, and of course far better as a diver, might have taken it down for the trial, but never before had we had the new torch deeper than 6 fathoms, while the depth here was 22. What effect would the deeper water have? None, I hoped; but in case it caused trouble or even another failure, I wanted to be on the spot to see for myself why. So, with the veteran Kelley for a helper on the bottom, I was rigged out in diving armor, and, following Kelley, for the first time I went overboard, to slide myself down the thin manila descending line to the bottom of the sea where so many divers had preceded me to struggle with that wreck.

Indelibly engraved on my memory is that first dive —the unearthly sense of loneliness as the sea swallowed me up; the dizziness as the pressure increased and the water flowed upward by me endlessly as I sank, always dropping, dropping into what seemed bottomless depths; the fierce cold that promptly cut through the woolen layers of my heavy underwear; the roaring of the air in my ears as it whistled into my helmet and gurgled out my exhaust valve to funnel upward in widening clouds through the water above; and the eerie effect of having suddenly entered a strange world, of seeing nothing at all pertaining to the world of men in the rapidly dimming grayness of the water about me except the manila descending line dissolving mysteriously into nothingness above while it materialized out of nothingness below, as, with arms and legs wound round it, down the line I slid.

When after having dropped through what seemed to me endless miles of ocean, the submarine itself finally loomed up beneath me against the dull background of the ocean floor, a silent hulk, massive and immobile, I lost all hope. Considerably magnified by the water through which I peered, she looked colossal in size, far beyond the power of human hands ever to raise. I had tackled the impossible.

But I had little time to speculate on that. The line to which I clung led not to the submarine itself but to a pontoon alongside its starboard quarter, and I slid on down past the *S-51's* conning tower to land in another moment atop the end of that pontoon, a gigantic cylinder as large almost as a section of the submarine itself, one of eight similar pontoons which, when attached to that submarine by chains beneath her keel, should tear her from her bed in the depths and drag her back to us on the surface.

I clung tightly to the descending line for support as I tried to balance myself on top of the pontoon, so dazed from the pressure that I felt each instant I would collapse, while a little distance away, Kelley's beckoning figure, magnified to at least twice life size, danced before my unsteady eyes. Falteringly I walked over to him, feeling that, when I crumpled up, I was safer at his feet than where I was. But Kelley, to whom all this was an old story, completely unaware of how close I was to unconsciousness, simply knelt on the pontoon and pointed to the link, lying across the pontoon hawsepipe, which he had twisted slightly so I could get at the iron stud which I was to burn out. Since my quivering knees seemed likely to buckle under me at any moment, I knelt beside him, my befuddled brain noting gratefully that now I should have less distance to drop when I folded up.

Again Kelley pointed to the stud in the middle of the chain link between us. Since I was still conscious, I concluded I might as well pass out working as idle, and, like an automaton, I dragged up the torch and its hoses, which were trailing behind me by a loop fastened to my wrist, subconsciously, I suppose, adjusted the flow of gases, and flashed a spark across the tip. The torch lighted instantly and continued to burn with a staccato roar like a machine gun firing continuously, setting up a pulsation in the water beating against my body which must violently have accelerated my circulation, for immediately my head began to clear.

I pressed the torch tip close to the 2-inch-thick square iron stud set in the middle of the chain link before me, held it there while the orange flame blazed against the metal. A few brief seconds, and then, glorious sight for my dizzy eyes, a spot on that iron stud glowed first a dull red, then brightened quickly into a limpid yellow. I pressed a trigger on the torch. A stream of pure oxygen shot out through the middle of the flame and hit the glowing iron. Instantly the miraculous happened. The flame, which had been at most a few inches long before, suddenly lengthened out to more than a yard, and there, streaming through the dark water like the trail of a rocket, was a dazzling array of white hot sparks, a torrent of burning metal more brilliant in its unearthly radiance in the dim ocean depths than any possible fireworks display!

Beneath the torch tip, framed by the whiteness of igniting iron, a black cut about an eighth of an inch wide appeared in the stud, and as I slowly drew the torch across, lengthened out till, perhaps a minute later when nearing the other edge of the stud, the cut was complete and the two severed halves of that heavy iron bar fell from the link with a metallic clatter to the pontoon beneath.

Deliriously intoxicated by the sight, I stared at the gap in that chain link where the stud had been. I had cut through a 2-inch bar of iron on the bottom of the sea, I had cut it on my first attempt, I had cut it starting in a semi-stupor! No doubt about it now—that torch was a success, it could cut steel regardless of depth, regardless of salt! My head cleared magically, as if that ocean load had abruptly been lifted from my body, for indeed a mental load at least equal to that crushing weight on my flesh had for the first time been lifted from my mind, and I could see success ahead. I never doubted again from that moment that we would lift that submarine!

We did, too, and in that achievement the underwater touch bore an important part, mainly in providing us with an easy method of securing the pontoons to the lifting chains. After that first success, Kelley did most of the cutting, though I myself took the torch down on two other occasions for burning jobs. The first of these was an emergency. We were all ready to unmoor in the face of a rising storm, when our last diver to come on board before we let go our lines informed us that the after mast on the submarine, a stubby steel affair which held the radio loop, was in imminent danger of puncturing the shell of one of our pontoons, and would certainly do it while the pontoon swayed to the storm waves, far below the surface though it was.

We had no other divers available, so I went overboard myself to cut off that mast, and a weird job it turned out to be. The deck was sloping sharply, there was no decent footing so far aft on the tapering submarine, and hanging on while I tried to burn was next to impossible. Clinging with one hand to the superstructure while I tried to guide the torch with the other, I could hardly help but make slow work of a job I could scarcely keep in touch with. To make matters worse, continually ringing in my ears were Lieutenant Hartley's admonitions that the storm had hit the *Falcon*, that her mooring lines were straining dangerously, that any instant he expected to be torn loose, and his urgings to me to quit and come up before it was too late.

Down where I was, I could hardly believe it. Around me was the solid ocean, calm, peaceful, and quiet—my only worry was to find something to hang on to while I burned on the low side of that mast. I doubt that I ever was in a more distracting position—clinging almost by my fingernails to the low side of a badly heeled submarine, one arm stretched far out holding a flaming torch in bare contact with her mast, and pounding continuously in my ears those anxious queries from the deck, ending finally in the peremptory statement that the ship could hold on no longer; in the gale lashing the surface the mooring lines might snap any second, they *must* let go!

So with only a slight amount of metal left uncut, a

strip so narrow that any minor blow would break the mast off, I reluctantly put out the torch, saw my life-lines clear of any wreckage, and signaled I was ready.

Immediately I was yanked up 40 feet, given a little decompression at the 90-foot depth, a little more at 60, and then, in one rapid lift, hoisted aboard the *Falcon* which was already casting off moorings. As I came up, the change from the quiet of the depths to the break-ing seas and the roaring of the gale while I was dragged through the surface and on board the plunging *Falcon*, was almost unbelievable. The waves broke continuously over the quarterdeck and drenched the dressers who were struggling to remove my weights and helmet. When these came off I was rushed forward into the recompression tank, to be put under pressure again; and hardly had I entered when, from the throbbing of the engine below, I could feel that we were already under-way and running for shelter.

One other episode with the torch on that submarine was especially gratifying. For various important rea-sons, we had to drop the anchor from the submarine, a mushroom anchor set in its keel forward. Because of wreckage, no diver could get access to the regular re-leasing gear inside the submarine to trip the anchor; still less could we reach the anchor cable at that housed anchor itself. There was just one chance.

The anchor cable was a heavy steel wire, which, on its way from the anchor windlass inside the submarine to the hawsepipe forward, passed through the super-structure in the bow of the submarine over the tor-pedo tubes, and there in the deck was a tiny square hatch over one of the sheaves which guided that wire. Through that hatch, a man might thrust his arm to oil the sheave (that was what the hatch was for) and in so doing he might touch the wire, but while in that position with his arm filling the tiny opening, he could hardly see the wire nor ordinarily could he do anything to it.

But that gave us our opportunity. If a diver could get his arm through that opening, he could get the torch through it. Kelley was rigged out and dropped overboard to try. He landed on the S-51's forecastle, walked forward, found the hatch, threw it back, looked in. It was dark inside that water-filled hole; he could see nothing. Adjusting his torch, he telephoned us to switch on the igniter. I did; Kelley lighted the torch. Sprawled out on the submarine's deck, Kelley thrust the torch through the opening, pushed his helmet forward over it. In its brilliant orange glow, he saw below a vague black streak running fore and aft—the anchor cable. Flattening himself on deck, he pushed the torch farther down till he could feel it touch the under side of the wire, held it there an instant, then pressed his trigger and slowly drew the torch upward. A shower of white sparks filled the opening below for a few seconds, then the wire parted. With a rumble and a roar that shook the submarine and was easily audible to me over Kelley's telephone, the cable snapped and the S-51's anchor dropped down the hawsepipe and fell clear.

Kelley telephoned then to shut off the gases and take up the torch. I looked at my watch. It was just

45 seconds since Kelley had lighted the torch till that
wire was burned through—a vast change from the time,
the autumn before, when the same Kelley, with teeth
chattering from cold, had angrily phoned up from the
bottom of the sea that if he could only get his teeth
onto another wire, he could chew it in half faster than
the old navy torch was burning it.

Since then, that torch has served on many jobs far
removed from submarines and from salvage, and even
from the sea. From the beds of rivers as far distant
from each other as the Arkansas and the Winooski, it
has served to cut up and remove twisted steel bridges
swept away by floods. On the seacoast, on one bridge
foundation job after another, it has changed the technique of construction, and on one complicated bridge
job in Louisiana it saved an immense foundation for
a bridge when the builders were hopelessly stalled.

The mellifluously named but rampaging Atchafalaya
River in flood washed away a railroad bridge spanning
it. The railroad, hastily throwing across a temporary
trestle and track near by, determined to install as a
new bridge a structure which no flood again could
damage, intending once and for all to halt the tie-ups
on its main line. But the engineers, in probing the river
bed, found to their astonishment that not till they got
down 185 feet below the river level was there a sub-
stratum solid enough to carry the load of the new
bridge.

There was nothing for it but to sink caissons to that
depth, unprecedented in engineering annals. So they

built a circular caisson over 20 feet in diameter, with
the bottom chamber sealed off by a heavy watertight
wood deck supported by 6 massive horizontal steel
girders 48 inches deep, to make an airtight chamber
in which the sand-hogs might work under compressed
air to remove the mud and quicksand below. They
started to sink the huge vertical cylinder down through
the river bottom. For this purpose, it was necessary to
keep the working chamber under air pressure heavy
enough to prevent the mud and sand from surging upward into it and drowning the sand-hogs; naturally as
the caisson went down this pressure had steadily to be
increased.

When the caisson had been sunk to 100 feet, the air
pressure reached the point where it was considered that
the sand-hogs could no longer stand it without undue
risk of fatalities from "the bends," and at this point
the technique of sinking the caisson was changed. Sand-
hog work was to be discontinued and the lower air
chamber was to be dismantled, with its girders unbolted
at their ends from the inner sides of the caisson and
from each other. The girders with the deck planking
forming the watertight seal were then to be dropped,
by releasing the air pressure beneath, into the pit below
last excavated by the sand-hogs. After this, the girders
were to be removed one by one by hauling them up
with cables already attached to them, and the caisson
was to be sunk the last 85 feet by excavating inside
it with clamshell buckets worked from the surface,
while high pressure streams of water, played through
fixed pipes already in place, jetted away the sand around

the lower edge of the caisson.

On the first caisson, on the west bank of the Atcha-falaya, the scheme worked as intended, and down through 185 feet of mud and sand went the vast cylinder, to come to rest finally on hard bottom. On the east bank, the second caisson went down also to the 100-foot level as intended, while the sand-hogs labored beneath it. At that point, with the decking overhead some 5 feet above the mud and 95 feet below the surface, the bolts in the girders were removed, the sand-hogs came up for the last time through the airlock to the surface, and the air pressure was released to let the girders drop.

But to the dismay of the engineers, the girders refused to drop into the pit below and clear themselves; in fact they failed to drop at all, and every effort to haul them clear by heaving with derricks on the lifting cables was futile. Why they wouldn't drop, heavy as they were, into the pit below them, was a mystery, but the reason for that was of no moment to the perplexed engineers. They had to get those girders out before the dredging could proceed and the caisson be sunk farther.

Securing a local diver, they sent him down through the 95 feet of water flooding the inside of the caisson to plant dynamite charges around the girders and blow them in two; but the charges had necessarily to be small, lest the explosions rupture the caisson walls themselves and thus completely ruin everything.

The attempt at dynamiting was a failure. All that resulted was that the girders were badly twisted, the

wood planking broken, the watertightness of the decking ruined beyond repair, and not a single girder was
cut. All were as tightly jammed as before.

Now indeed the job was stymied. It was no longer
possible to put an air pressure in the lower chamber,
expel the water from it, and send the sand-hogs down
again to clear away whatever was preventing those refractory girders from falling. Nor, with that mass of
tangled steel blocking off the inside of the caisson 95
feet down, was it possible to drop the clamshell buckets
to the mud below and continue the excavation necessary to sink the caisson farther. The caisson was worse
than useless. Only halfway down to hard bottom, it
could carry no load, but there, directly in the path of
the semi-finished bridge, with the bridge trusses, already
fabricated, waiting to go up, it effectually prevented
the proper placing even of a substitute caisson and definitely blocked the completion of the bridge. Meanwhile, creeping slowly by on the swaying temporary
wooden trestle, the heavy trains rolled while, in distant railroad offices, every executive from president
down to division superintendent, with the main line
tied in a Gordian knot, chafed at the delay and swore
at the luckless bridge engineers.

John Kelley cut the knot for them. Rushed down
posthaste from New York at the frantic request of
one of the bridge engineers, who had read that something new in the world had helped salvage a submarine
and trusted that it might also help pull a railroad out
of a desperate hole, far from the salty sea where first he
had watched me as a novice diver cutting steel, Kel-

ley turned to.

Amidst a terrifying tangle of wire, broken planking, and twisted steel left by the dynamite explosions, Kelley groped in the blackness beneath 95 feet of Atchafalaya mud and water to get at those girders with the torch. Diving twice a day for a total of 18 hours under water, he cleared away the debris, located the 4-foot deep girders, burned through the thick steel to cut all 6 of them in half, hooked new cables to the severed pieces, and then, on the topside, one by one watched the derricks hoist the bent sections well above the top of the caisson and drop them into the mud outside. Hardly had the last section come clear of the inside of that caisson, when the clamshell buckets began to splash into its interior and the dredging commenced. My underwater torch had saved that bridge.

That was eleven years ago. It is hardly likely that any passenger, rolling westward from Baton Rouge, as his train rumbles at full speed over that massive bridge across the Atchafalaya, realizes that in part he owes his steady and safe crossing to a diver working far below that river with a flame which first flared out successfully on a shattered submarine in the depths of the distant Atlantic.

STEEL GIRDER FROM A WRECKED BRIDGE, BURNED IN HALF ON THE BOTTOM WITH THE ELLSBERG UNDERWATER TORCH

CHAPTER XI

Sunken treasure!

There is a lure in those words that is older than history, that has dragged men to the ends of the earth and to the depths of the sea, a lure that has broken the health of many adventurers, cost the lives of many divers, and stripped the last copper from the purses of thousands who have put up the cash for no one knows how many salvage expeditions down through the ages. And it still goes on.

Poets are as responsible as anyone for the queer state of the public mind on this subject. Shakespeare echoed what must have been the popular conception of his day when he made one of his characters—who, in a dream, thinks he has been knocked into the sea and as he drowns observes the depths—say:

> Methought I saw a thousand fearful wrecks;
> Ten thousand men that fishes gnaw'd upon;
> Wedges of gold, great anchors, heaps of pearl,
> Inestimable stones, unvalued jewels,
> All scattered in the bottom of the sea.
> Some lay in dead men's skulls; and in those holes
> Where eyes did once inhabit, there were crept,
> As 'twere in scorn of eyes, reflecting gems,
> Which woo'd the slimy bottom of the deep,
> And mock'd the dead bones that lay scattered by.
> (*Richard III*, Act I, Scene IV.)

193

In this opinion, Shakespeare was neither first nor last; nothing in this picture has changed in the centuries since his day save that in the lurid circulars put out by so-called salvagers to entice the unwary, the wedges of gold have so multiplied as to blot from sight the great anchors, and the skulls are no longer visible through the enlarged heaps of gems which now completely engulf them.

If one wishes only to believe, there is ample legend to go on. Every Spanish galleon driving eastward through the Gulf of Mexico was nearly awash amidships because of the vast treasure in gold, in silver, and in gems which crowded her hold; only the towering poop and the forecastle, protruding fantastically from her hull, provided the buoyancy to keep her afloat under her load of treasure. That the West Indian islands, Cuba, Haiti, Puerto Rico or Jamaica, had so little gold and silver that Columbus, their discoverer, died in poverty, a suppliant before the throne, is quite beside the point. All proper Spanish galleons, regardless of port of departure, were always laden with treasure; and consequently when, due to scanty charts, poor navigation, storms, or poor construction, any were wrecked or foundered at sea, the volume of sunken treasure was immediately increased.

Common sense would indicate that these islands at least, which provided a large part of the commerce across the Spanish Main, could never have furnished cargoes that class as treasure. It was the bitter disappointment of Columbus' aging years that he found there in four heart-racking voyages nothing of the civilization of Cipango which he was seeking, none of the wealth. Nothing was there except naked savages and a tropical wilderness; he died in the belief that, while he had found a new route to the East, he had not yet really come upon the islands of spice, of silks, and of fabulous eastern wealth.

But did that stop the legend makers? One of the best treasure lures of history concerns Columbus' own time and indeed Columbus himself.

So disappointing were the actualities encountered in the new-found "Indies" that among the colonists brought out by Columbus, and responsible to him as viceroy, bitter controversies verging on mutiny arose and found their way across the ocean to the court of Ferdinand and Isabella. As a consequence, a new viceroy, Francisco de Bobadilla, was appointed in his place, and, in the year 1500, set out for Hispaniola with full power to supersede the discoverer and to conduct an investigation into his conduct.

Bobadilla, pompous, inefficient, and arbitrary, the perfect picture of the bureaucrat in any age and in any land, promptly upon arrival in Hispaniola ordered Columbus before him, and, having read his royal warrant as authority for his conduct, ordered the discoverer of the New World put in irons for transportation as a prisoner back to Spain. Amidst the clanging of hammers, the irons were attached, and, loaded hand and foot with chains, Columbus was sent home as a felon.

The very severity of Bobadilla's course defeated his purpose. That Columbus had found little gold was evident; that he might have mismanaged his colony was possible; but upon arrival the sight of the old man in fetters sent a wave of indignation seething across Spain that made unnecessary the slightest defense on his part. Freed immediately, he was received at court with every mark of consideration, and the kind-hearted Isabella, on greeting him, wept openly over his indignities. Restored to his rank and most of his authority, he was soon outfitted once more with another fleet and set sail again on his fourth and last voyage to continue his discoveries, the sole restriction on his movements being that the shrewd Ferdinand forbade him again to visit Hispaniola, the scene of his late difficulties.

Meanwhile Bobadilla was busily engaged in getting himself into hot water. His own gross mismanagement of affairs, his cruelty to colonists and natives alike, and his insufferable insolence so enraged the colonists of Hispaniola that the stories filtering back to Spain, made easily believable there by his conduct toward his predecessor, resulted in his being himself recalled under arrest to render an account of his stewardship.

Now enters legend. To soften the wrath of their Most Catholic Majesties, of which he had much reason to be afraid, it is said that Bobadilla gathered up all the gold in the island, and, in the holy city of Santo Domingo, had it cast into a magnificent golden table weighing over 3,000 pounds—by present-day standards worth well over $1,600,000. With this as a present for his monarchs, he might well hope for honors at the court, instead of the merited chastisement, and, considering the later treatment of Pizarro, Cortez, and Henry Morgan at the courts of both Spain and England, he may well have forecast the royal psychology of that day—gold excuses all.

And now back to historical fact. With a fleet under the command of Torres, the suspended governor prepared to sail for home in accordance with his peremptory orders to return forthwith. At this juncture, by a freak of fortune, there arrived off Santo Domingo the squadron of Bobadilla's late victim, commanded by none other than the Admiral of Ocean Seas himself!

Disregarding his royal orders to keep clear of Hispaniola, there was Columbus again off its capital. But he had ample reasons for his conduct. His vessels were in bad shape and in imperative need of a refit before they proceeded on their scheduled voyage of discovery, and nowhere else in the New World in 1502 were there such facilities for repair as in Santo Domingo.

Lying to outside the harbor, Columbus sent word in to the viceroy asking permission to enter for repairs, giving as an additional reason his belief that his ships were in grave danger from a storm he believed to be imminent. But the vindictive Bobadilla summarily warned him to stay outside. With no alternative, Columbus with his own meager facilities refitted his vessels as best he could while they lay off the coast and then prepared to proceed, but, learning meanwhile that Bobadilla was about to weigh anchor for home, sent him another message begging him to delay his departure for a few days as a fierce storm was surely due. Bobadilla haughtily rejected his advice and made sail with his fleet eastward, while Columbus, commending himself, no doubt, to the hands of God, sadly watched the safe harbor of Santo Domingo fade astern as his own fleet stood westward along the south coast of Hispaniola.

Two days later, the storm broke, a raging West Indian hurricane. Columbus, proving his title of Admiral of the Ocean Seas, by consummate seamanship weathered it in his flimsy ships by hugging the south coast of Hispaniola, for him a weather shore affording some protection, but with Bobadilla it was different. Somewhere to the eastward, believed to be off the north coast of Puerto Rico, and for him a lee shore, the hurricane scattered his fleet, wrecking Bobadilla's ship and drowning Bobadilla. And the golden table? That, of course, was lost too, to become one of the fantastic treasure legends of the deep for which divers may seek. Where? Somewhere off Puerto Rico, for certainly neither Bobadilla nor any member of his crew, driving before the indescribable fury that West Indian hurricanes develop, could have had any very definite idea of where they were when their galleon sank, nor were they at the moment, with death staring them in the face, particularly interested, even had they known, in marking any charts with crosses for the benefit of posterity.

Did such a table ever exist? It is hardly likely, considering the barbarous state, in 1502, of the natives in the islands discovered by the Spaniards, and their lack of civilization, cities, temples, and mines. But that has not prevented the fabulous golden table from reaching a weight of over one and a half tons, to become a lure for treasure-seeking divers with more imagination than experience and with contraptions as fantastic and as unhampered by reality as the treasure which they seek.

CHAPTER XII

STILL, it is unnecessary to go back to the days of the Conquistadores for tales of sunken Spanish treasure—there is at least one modern case which has all the necessary embellishments for treasure seekers, and a history almost as unbelievable. Bullion, gems, romance—all are there. What more could one ask before the divers set forth to dare the deep?

Early in May, 1911, the S.S. *Merida* sailed from the harbor of Vera Cruz, Mexico, for New York. She was a prosaic coastal liner, not overly large, never the type that, for luxury, speed or importance of passengers, would make the news. On May 12, 1911, while well off the Virginia Capes, northward bound in a light fog, she was in collision with an even more prosaic vessel, the freighter *Admiral Farragut*, which crashed into her side, inflicting a mortal gash.

That was about midnight. As the *Merida* took over five hours to sink, her captain had little difficulty in transferring first his passengers and then all his crew to the *Admiral Farragut*, and, before he finally abandoned ship, had time even to set his stewards rummaging through the passengers' cabins to save what effects seemed worth saving. Not till 5:20 A.M., with his vessel down by the stern and badly listed, did he finally leave, to watch his ship sink some ten minutes later, in about 200 feet of water.

That was less than thirty years ago. Under ordinary circumstances, the *Merida* would simply have joined the long roll of other wrecks and, as soon as the underwriters had paid the losses, been quickly forgotten; but the circumstances surrounding the *Merida's* sailing had not been ordinary.

For some time before, Mexico had been torn by civil war, which for Mexico was nothing novel; but with this revolution, unlike many before in his long reign, Porfirio Diaz, aging perpetual president of our sister republic, was unable to cope in his accustomed manner. Accordingly, he decided to abdicate while still he had sufficient control of his army and the government to make graceful abdication possible, and departed for Paris, leaving as his almost certain successor the head of the revolutionary party, Francisco Madero. Diaz, foreseeing such a possibility, had long before prudently transferred to France a considerable part of his fortune —considerably more he took with him—and ended his days in Paris a wealthy *émigré*, little concerned over the continual turmoil in Mexico which followed the end of his long dictatorship, as successive patriots fought for the presidency and, not so fortunate as Diaz, usually left office in coffins.

It was during the very month of his abdication, but some weeks before that event, that the *Merida* sailed on her last voyage. Conditions in war-torn Mexico had for many months before been, to say the least, unsettled, with much of its territory ravaged both by the contending armies and by roving bandits, a condition which

lasted for years after. Political refugees were common
along the American side of the border and in New
York. They had as their outstanding characteristics the
twin facts that they were against the Mexican government of the moment and that they were broke.

Under such political uncertainties, then, the *Merida*
weighed anchor and left Vera Cruz, carrying among
her passengers the usual quota of refugees, some des-
tined on arrival at New York to sponge in the accustomed manner off their compatriots and any others with
a few dollars to give in exchange for the somewhat
threadbare tales of how their vast *haciendas* had been
ravaged or confiscated, and they themselves reduced to
beggary. But chance gave to the *Merida* voyagers a far
better story. Beyond any contradiction, their ship had
been sunk; they landed as shipwrecked mariners who
had lost their all at sea.

And what an all it finally turned out to be! Imagination ran riot. The crown jewels of the Emperor Maximilian (whom Profirio Diaz half a century before had
helped to dethrone and execute) and of his Empress
Carlotta; the treasury of Diaz himself; the wealth of
Mexico in gold and silver—all these and more were
traveling northward in the baggage of the fleeing
émigrés on the *Merida*. Undeclared, of course unin-
sured, secretly packed to avoid seizure by, of all persons, the very agents of Diaz who still held Vera Cruz
when the *émigrés* departed! Odd, certainly, that most
of the refugees on previous and on following ships
had managed to escape from Mexico only with their
lives and with a few personal belongings, and were

thankful for small loans, while these, their compatriots
on the *Merida*, had it not been for an unfortunate accident at sea, would have landed with the wherewithal
not only to have outshined Solomon in all his glory,
but to have financed a counter-revolution in Mexico
which would have restored every refugee to his lost
hacienda. On the strength of these sad tales, undoubtedly many a sympathetic listener lent aid to the dis-
tressed victim of the ravening waters.

What are the facts about the *Merida's* cargo? Not
long ago, in the offices of the admiralty lawyers who
handled the claims for the underwriters and settled
them, I went over the losses and claims therefor. There
were claims for the ship and for the insured cargo and
personal effects directly against the underwriters who
had carried the risks. There were claims against the
Admiral Farragut for having caused the accident; and
against her any claimant might have lodged a libel for
his losses, regardless of whether his goods were insured
or not. Certainly any passenger with the slightest scin-
tilla of evidence to establish his loss would at least have
filed a claim for it, for, with the wreck fresh in the
records, such evidence might well have been obtained,
even in Mexico, did it exist. And had the crown jewels
been missing from Mexico and aboard the *Merida*, certainly the Madero government, then the legal owner
of them, regardless of how they came aboard the
Merida or in whose hands they were traveling, would
have lodged a claim for their value.

What happened is illuminating. The passengers filed
claims for personal belongings to a total of $168,000.

The crew claimed a loss of $8,000 for their effects. There was a claim on cargo for around $400,000. Of the cargo, $240,197 worth consisted of silver bars or bullion, and $26,730 worth consisted of copper; the remainder was general merchandise, hides, etc. Finally there was the claim for the loss of the *Merida* herself, by far the most expensive item. The Mexican government filed no claim at all, apparently unaware that it had lost anything.

Having all the claims before them, the underwriters proceeded to settle, paying the insured losses without discussion, the others by compromise, and promptly wrote the *Merida* off their books. They were out of pocket for the losses and were now the legal owners of the *Merida* and all her contents, had they any desire to recoup through salvage, but neither then nor since have they ever spent a dollar in any effort at recovery. There was nothing aboard her worth salvaging that they knew about, save a moderate quantity of silver, some $240,000 worth, and, as prudent business men, they figured that the cost of salvaging that silver was not warranted by the recoverable value. So far as the millions aboard her in the way of gold and jewels were concerned, they assessed those legends as moonshine, and proceeded to forget the *Merida*.

But others with more romance in their souls did not. A few years went by, just enough to give the stories time for proper mellowing and growth. The rubies in the crown jewels grew to enormous size; the hoard of gold increased proportionately. A syndicate was formed, composed astonishingly enough of half a dozen men whose names bulked large in finance in Philadelphia and New York. Certainly to them the possible rewards were not important; only the romance of the enterprise could have really mattered.

In the most approved manner, the expedition proceeded. A small vessel was obtained, fitted out for diving. Two divers were procured, on monthly terms which would make the average diver's mouth water. Wreckmaster and crew matched all else. The expedition sailed to salvage the *Merida's* treasure.

For two successive seasons, this expedition worked on the *Merida*. What went on has never been given much publicity. As is usual in such cases, the wreck was located near the end of the first season's work, thus giving the backers of the expedition something to dream over during the long winter and to keep them in a properly generous frame of mind till spring. The second season was similarly barren of any results in the way of gems and bullion or anything else of a concrete nature. It ended with all the funds subscribed originally by the syndicate expended, but with the most glowing tales of what might be expected in the way of gold and jewels to be brought up during the third season.

At this point, however, the wealthy syndicate members, having acquired a sufficient amount of romance to last them for the remainder of their lives (and perhaps having begun vaguely to fear that they, instead of the *Merida*, were the real treasure vessel), refused to put up any more cash. With this decision, that div-

ing expedition over the *Merida* came to an abrupt conclusion.

But the *Merida* legend did not die. Soon another
group of treasure seekers were in the field, financed
this time by stock subscriptions, with a weird contraption that would make divers and diving unnecessary
and quite easily harvest the gold and jewels in the
wreck. A submarine tractor, fitted out with gadgets
enough to do most anything, was to travel along the
ocean floor carrying an operator inside to work the
intricate assembly of drills and hooks protruding from
it, and prepared to do whatever was necessary to the
defenseless *Merida*. In this case, the idea was to spread
a huge canvas on the ocean floor alongside the ship,
blow her up with TNT planted by the tractor, and
then catch all the gold and jewels as they rained down
through the water on the canvas. After this nothing
further was required save to lift the canvas by the corners to the surface and take aboard the treasure.

This scheme, startling in its simplicity, eliminated
all the hazards, the delays, and the uncertainties of
working with divers in deep water. The promoters
sold stock enough to build their tractor, something very
much resembling a baby tank, but far heavier, as it
weighed in the neighborhood of 10 to 15 tons. Before
going farther in the deep sea, and possibly to attract
more investors, they proceeded to stage a demonstration of the tank's abilities on a small steel vessel which
they sank "inadvertently" in Long Island Sound in

rather shallow water not far offshore. So far, all went well; newspapers even carried a story where the sinking was reported as a bona fide accident, with the new-fangled tractor called in to salvage the wreck. All of this publicity no doubt made the promoters rub their hands gleefully.

However, from this point on, matters went awry. The submarine tractor, rushed to the scene on a floating derrick (the promoters had difficulty in hiring one husky enough to handle it) was lowered to the bottom; but instead of rolling over to the side of the wreck and getting to work with its drills and hooks, it ingloriously refused to move and had to be heaved up again by the derrick, with nothing done. It appeared that, by a curious oversight of nature, the bottom of the Sound was not properly macadamized, and the inordinately heavy tractor found itself bogged down and incapable of locomotion. The tractor demonstration was a failure.

Then to make matters worse, a heartless government insisted upon the removal of the "wreck" as an obstruction to navigation, and the unlucky promoters found themselves obliged to call in a commercial salvage company and its prosaic divers to get lines attached to the little vessel so it could be carried away.

This closed the active career of the tractor, which never actually got anywhere near the sunken *Merida*. The tractor, I suppose, has long since gone to the junk pile, which is possibly just as well. Had it ever succeeded in getting alongside the *Merida* and planted its charge of TNT there, the canvas might have proved too small to catch all the treasure tossed out by the explosion. On the other hand it is possible that it might have caught nothing but scrap iron and thus have ruined the *Merida* legend for future treasure seekers.

The next *Merida* phase was more exciting. Another inventor turned up with a trick diving suit, one of the "armored" variety intended to protect the diver from the water pressure and thus save him from "the bends" and all similar diving complications. That incidentally the diver, locked up in that contraption, with nothing but protruding hooks for hands, was rendered practically incapable of doing any useful work or of getting about, was a minor matter. The suit was complicated enough and terrifying enough in appearance to give it a definite sales appeal, and the usual stock promotion followed. This time the *Merida's* treasure was certainly going to be salvaged.

Unfortunately the sales of stock proved slower than anticipated, and not for several years after the announced date did the salvage expedition sail for the waters off Cape Henry. But then the salvagers proved themselves showmen of the first water and made up for lost time. In front page stories all over the country, one morning the news broke in wireless reports that not alone one but *two* salvage expeditions were simultaneously after the fabulous treasure of the *Merida*, savagely battling each other for the privilege of mooring over the wreck and getting their divers down to fetch up the gold and the crown jewels. Maximilian, of whom most Americans had never heard, became a breakfast table topic, and the tragic fate of the Empress Carlotta was intertwined in the news with fierce radio accounts of severed anchor cables, slashed mooring lines, and threats of bloody reprisals as the rival salvage vessels desperately maneuvered for diving positions over the coveted treasure. Finally after a week of turmoil, an armed Coast Guard cutter was sent out with orders to prevent bloodshed and at any cost to keep the peace on the high seas.

In the face of this show of force, the two competitors decided to compose their differences and coöperate, leaving cynical salvage men to wonder whether it had not been all a plant, with both vessels under one control, and with the battle staged solely for publicity purposes, in which aim the promoters apparently succeeded beyond their wildest expectations.

What happened after that was, so far as the treasure seekers were concerned, shrouded in deep silence as compared with the fanfare in which the expedition began. Nothing was accomplished, nothing was recovered, in spite of their impressive metal armored diving rig which was to make the recovery of the *Merida's* treasure simple beyond words. (So simple in fact that that same summer, the promoters assured their investors, they would also salvage the *Lusitania*, the *Egypt*, and then move on to spend the winter salvaging more treasure from Navarino Bay in the Mediterranean.) Only once more did any of these embattled treasure seekers burst into print over the *Merida*. Not long after, one

of their diving ships came into Norfolk with its chief
diver dying of "the bends," and the expedition folded
up.

That left the *Merida* to rest in peace for a few years
more. But only till 1938. For that year, lured by the
stories of the *Merida's* treasure, which certainly had
lost nothing in crossing the ocean, a hitherto unknown
group of Italians decided to try their hand on her.
From Spezia, Italy, their departure in the *Falco*, a small
steamer, was announced in late May against the usual
rosy background of stories about the jewels and the
gold, amounting now to the figure of $4,000,000.
Twelve days later they arrived at Norfolk, where once
more the crown jewels and the Emperor Maximilian
were paraded in the news. The *Falco* and her crew
then sailed for the scene of the wreck, leaving the unfortunate impression that they were the Italian group
which had successfully conducted the salvage of the
Egypt (which most decidedly they were not). Since
then a pall of silence has once more fallen on the
Merida. It is known only that the salvagers returned
to Italy with nothing at all of value recovered from
her.

A competently directed and manned salvage expedition would have a fair chance of blasting and cutting
its way into the *Merida's* hull and recovering the silver
there. More difficult salvage jobs than this have been
successful. But in 200 feet of water on a heeled-over

wreck, the task is both difficult and dangerous, and no job for amateur romance seekers or stock-selling promoters and the type of salvage "experts" who usually travel in their train. Real divers and real salvage masters could do it, but a properly organized and fitted out expedition is expensive and deep diving operations on a difficult wreck are always long-drawn-out and costly (as will be shown later). For the known treasure, $240,-000 in silver, no competent American authority would recommend diving, because the cost of a proper expedition over two seasons might well exceed the value of the silver when recovered, and there is no guarantee of success even in two seasons' work. For a low cost operation, such as might be run by the Italian salvagers of the *Egypt*, with a much depreciated lira and a much lower wage scale than prevails in America, $240,000 might prove moderately profitable if successful, but even with them the real prize is too small to warrant the risk with those who really know salvage.

So the *Merida* rests at the bottom of the sea. In spite of reason, in spite of the well-authenticated facts about her cargo, she is still surrounded by a halo of romance and loaded to the gunwales with nonexistent gold and enormous rubies, the perfect symbol of the sunken treasure legend.

If such tales can gather about an ordinary coastal liner, sunk less than thirty years ago, how much faith can be put in the glowing treasure legends of lost Spanish galleons in the great days of the Spanish Main, where three centuries or more have gone by, making it impossible to establish the facts while giving ample time for rumors to acquire a satisfactory embellishment?

To put the final touch to the comic opera atmosphere surrounding the *Merida's* treasure, the crown jewels of Maximilian and Carlotta at last burst legitimately into the news in late 1938. A Mexican, seriously desirous of acquiring them, instead of wasting time and effort and the risk of "the bends" diving on the *Merida,* took the much simpler and more sensible course of going where they were—to the National Museum at Mexico City where for years they have been publicly exhibited.

What then occurred, as reported by the *New York Times,* follows:

Stole Carlotta's Jewels

Laredo, Texas, Oct. 4, 1938 (A.P.). Theft of the crown jewels which once adorned the Empress Carlotta when Emperor Maximilian ruled Mexico was solved today by Federal officers.

Enrique Mazando Cordova, 23 years old, who was arrested today on a charge of smuggling, calmly admitted the theft of the crown jewels, and deliberately related how he engineered the daring plan. The jewels were stolen on Sept. 4 from the Mexican National Museum.

Cordova pleaded guilty to the smuggling charge before United States Commissioner Frank V. Hill and was held for the Federal grand jury in $3,000 bond.

The prisoner said he entered the Mexican National Museum on Sept. 4 and concealed himself. He waited until the closing hour, and when the building was deserted, he said, he walked to the large show case where the jewels were kept, took them, and left by jumping from a balcony.

For thus directing attention to the facts in this matter, a vote of thanks is due Enrique Mazando Cordova from all future "investors" in salvage schemes on the *Merida*. As soon as the smuggled jewels have been restored to the Museum for exhibition, they might think it wise in part to follow his example, buy a ticket to Mexico City, examine at their leisure the crown jewels, and decide after that inspection whether they are valuable enough to warrant risking any money in diving operations to recover them from the wreck of the *Merida*.

CHAPTER XIII

R.M.S. *Lusitania* is another star of the first magnitude in salvage promotion literature. Unlike the *Merida*, the *Lusitania* has always been news in America, from the days of her first voyage when, due to disgruntled stokers, she failed to capture the blue ribbon of the Atlantic, through her second voyage when she achieved the title of speed queen of the Atlantic, down to the fatal hour on May 7, 1915, when a torpedo from the *U-20* sent her to the bottom and she ended her career in a blaze of publicity never before or since equaled by any ship. Naturally enough, with the *Lusitania* a household word both here and abroad, she qualifies perfectly as a treasure ship for gulling the public.

A little intelligence in this matter is however worth while before engaging the divers to struggle in the 240 feet, more or less, of water in which that Cunarder lies. Did the *Lusitania* have any treasure aboard to dive for?

Transatlantic liners do carry gold. Never in the history of the world have such huge gold shipments across the seas taken place as occurred early in the World War, and, especially, in the last few years since Britain went off the gold standard. Financial unsettlement abroad and recent fears of war in Europe have sent cargoes of gold scurrying across the Atlantic in undreamed-of quantities. During the month of September,

1938, from England alone the shipments amounted to $370,000,000, and in the two months of September and October, 1938, a billion dollars in gold flowed into the United States from abroad, in cargoes that dwarfed into insignificance even those of the days when the loot of Peru and Mexico was flowing toward Spain. Single shipments reached as high as $50,000,000; cargoes from $5,000,000 to $25,000,000 were common; and even third class passenger ships, too slow ordinarily to obtain any of this profitable cargo, became modern argosies laden with gold bullion running into millions.

It is well to note that during this vast movement, even with a war brewing and liable any day to burst into a major conflagration involving practically all Europe, there was no attempt at concealment of commercial shipments. The amounts shipped each day and the vessels carrying the bullion were daily reported (as they have always been) in the press both in London and New York; the arrivals received similar attention. And of course each shipment was insured. Not for over a hundred years, in war or peace, has any considerable commercial shipment of bullion moved across the Atlantic without that protection.

With one exception (later considered) there have been no losses by shipwreck since 1914 in all this ebb and flow of gold between England and America. But does anyone think that, should one of these treasure-laden ships be sunk, the slightest doubt would exist in the minds of the owners, or of the underwriters, as to what the ship was really carrying in the way of unusually valuable cargo? This knowledge, the first essential in any bona fide salvage attempt, is available for all modern wrecks, and a reasonable amount of research will disclose equivalent data on older catastrophes. Yet in the case of practically every vessel which forms the object of the average salvage promoter of the present day, nothing authentic can be shown proving that the ship carried anything resembling in the slightest her supposed treasure. The reason obviously enough is that where a real loss has occurred, making salvage worth while, the underwriters always contract for salvage themselves, and, oddly enough, are usually successful. It is only the worthless wreck that is left lying unclaimed, to be used as a bait for the unwary. The *Lusitania* is a case in point.

The *Lusitania*, sailing from New York for Liverpool near the end of the first year of the World War, had the eyes of all riveted on her by the cryptic warning to travelers published by the German Embassy before she left. So before departure she was carefully examined by the port and customs authorities in New York. She had not the slightest chance of concealing anything in her cargo, even had the owners or shippers so desired. She sailed carrying a distinguished passenger list of Americans who, knowing they were sailing on an unarmed merchant ship, felt safe from enemy attack. But unfortunately for her passengers, she was, without warning, the target of a German submarine and went down on May 7, 1915, off the southeast coast of Ireland, carrying to their deaths 1,198 persons, including 124 Americans.

While there had been nothing secret about the *Lusitania*, her cargo, or her movements up to the moment she disappeared beneath the waves, immediately thereafter she became enmeshed in a cloud of legends, with the easily obtainable facts deliberately obscured by a flood of propaganda intended to justify her destruction. Nothing was too wild to be believed about the *Lusitania* —she was an armed ship, she was loaded with explosives, and naturally she was stuffed with gold.

The rumors that had to do with her character as an unarmed merchant vessel, prohibited from carrying explosives if she carried passengers, were matters of grave concern to the United States Government, intent on maintaining its neutrality and safeguarding its citizens. The rumors were proved to be wholly false; the *Lusitania* carried no explosives, she carried no guns. The propagator of at least one of the stories of her carrying artillery was convicted of perjury and sentenced to prison. But neither our Government nor that of Great Britain has ever bothered to go to similar lengths in squelching the equally imaginative stories of the treasure on the *Lusitania*.

In spite of the absurdity of such yarns, treasure-hunting expeditions on the *Lusitania's* hulk have tried to raise money from the public both here and in England. Here is a vessel which has been sunk hardly twenty years; a vessel which the owners, the underwriters, and the British Government all deny carried any appreciable amount of gold; a vessel whose position as she lies on the bottom is known within reasonable limits, and on which in 240 feet of water, salvage work could be done if it were warranted.

Yet not a stroke of salvage work has ever been attempted by the owners, the underwriters, or the British Government, thus proving the sincerity of those in the best position to know that there is nothing aboard worth salvaging. Still the *Lusitania* is usable bait for promoters with novel salvage gadgets and a well-founded faith in the ability of the public to swallow anything.

The *Lusitania*, when she sailed from New York, carried a cargo totaling $735,000 in value, listed in considerable detail in her manifest which was open to inspection. It is a curious list for investors in treasure-seeking schemes to ponder:

5,470 boxes of unloaded brass cartridge cases (no powder)	$200,000
Sheet brass	49,565
Copper (mostly wire)	53,000
Cheese	33,000
Furs	119,000
Silverware	7,000
Precious stones	13,350
Miscellaneous cargo	260,085
TOTAL	$735,000

The salvage value of the furs and of the cheese needs no discussion. The ammunition cases, the sheet brass, and the copper wire, for twenty years submerged in salt water, would now, as junk, command only a trifling part of their original manufactured value. The miscellaneous cargo would not be worth hauling up. So the treasure of the *Lusitania* comes down to some $20,350

in silverware and precious stones, plus what little in
personal effects might be obtained from the purser's
safe. The reasons why those financially involved, and
who initially had to stand the losses, have never bothered with salvage on the *Lusitania* become immediately
obvious.

Even if one wants to believe that the *Lusitania's*
manifest was falsely sworn to (though only Heaven
knows why, for the ammunition cases she was carrying were all listed in great detail and openly), that then
and now, the British Government for war reasons is
trying to conceal the facts surrounding the sinking and
has therefore discouraged salvage work—still, slight consideration of the circumstances will show the impossibility of any large amount of gold aboard.

For the flow of gold between England and America
is like a tide governed by inexorable economic conditions which dictate the direction of its flow. When
financial conditions are such as to cause gold movements in quantity, then during that period the shipments are always in one direction. No more can the
golden tides flow in both directions at once than can
those of the ocean, which under the compulsion of
the moon and the sun follow in the direction of
heavenly forces dragging them on.

During the early days of the World War, the flow
of gold was eastward as Britain and her citizens called
home their credits in America, seriously dislocating in-
ternational exchange and forcing gold shipments to
London to balance the account. But as 1915 dawned,
it became increasingly evident that the British Empire

was faced with a long war; American raw materials—
cotton, wheat, steel, copper—and such munitions as
could be obtained were badly needed in England,
which started to purchase here on an increasingly vast
scale. Contrary to present popular belief, a large part
of this material was actually paid for up to the time
we entered the war ourselves, much of it by mobiliz-
ing American securities abroad and returning them to
us in payment, some of it by payment in sterling bills
of exchange. To keep the British pound sterling at a
point where its purchasing value here was high became
almost as important a matter to England as holding the
Germans on the Western Front. When necessary to
bolster the value of sterling exchange, they shipped
gold westward to New York. The flow of gold to
London had stopped long before the summer of 1915,
and the tide was setting strongly westward. Unfortunately for the treasure legends of the *Lusitania*, she was
headed *eastward* when torpedoed. There was no gold
in her.

But might not the *Lusitania* be herself salvaged, lifted
to the surface for her value as a ship? After all, when
she went down, she was the pride of the Atlantic. Why
not have divers do that, as some amateur salvagers
suggest?

Why not, indeed? Because that is quite as ridiculous
as salvaging her for her mythical treasure. The *Lusitania*
was a vessel of about 35,000 tons. She went down in
the exposed sea some 10 miles offshore, in a terribly
damaged condition from torpedo explosion, and was
unquestionably further damaged by impact with the

sea floor. What does raising the ship herself now mean?

To raise her as a dead weight may be instantly dismissed. No surface lifting gear capable of dragging up 35,000 tons exists, nor, if it did, could it safely work in the open sea. A lift by attaching to her a multitude of submersible pontoons is remotely conceivable, but, to one who has struggled with the diving job of lifting a 1,200-ton submarine, is thoroughly foolish in this case. As for making the *Lusitania* lift herself by sealing up all openings in her hull and expelling the water with compressed air, as submarines are sometimes lifted, this could only be done in the highly improbable case that she lies upside down on the ocean floor, so that, bottom up, she has made of herself a huge diving bell. Right side up or on her side, which is more probable, the innumerable hatches and smokestack openings to be sealed make it totally impracticable to hold compressed air in her hull.

Upside down, in sheltered waters around 100 feet deep, some large ships have been raised that way, notably the ex-German battleships which were scuttled by their crews at Scapa Flow and capsized in sinking because of their enormous top weights in turrets and guns. These ships, still upside down, were raised by British divers, a most notable salvage feat. But the *Lusitania* is not in sheltered waters, she is beyond reasonable depth for long-continued *work* by divers, and she is known to be so badly torn open by the explosion which sank her that, regardless of her position on the bottom, she cannot reasonably be sealed up.

Finally, while the lifting of the *Lusitania* is technically not wholly impossible, the cost would certainly exceed that of building a new and modern *Lusitania*, and to what end? The *Lusitania* has been only junk from the moment she sank. Even at the high junk price of $15 per ton at the melting furnaces, the gross return for raising her as a ship would be only $525,000, a sum which would not even start to cover the cost of lifting her. To add the final touch needed to make any thought of lifting the *Lusitania* ridiculous, her sister the *Mauretania*, still in perfect running order and undamaged, was so poorly regarded by the Cunard Line as a ship that they sold her for junk several years ago and sent her on her last voyage to the shipbreaker's yard.

Briefly then, the *Lusitania* as an object for divers to work on can only be classed as a fraud. There is no treasure in her, and there never was. As a ship, no diver who belongs outside a lunatic asylum would ever wet a suit endeavoring to lift her.

What are the stories in the press, over the radio, about salvaging the *Lusitania* all about then? Three years ago a syndicated series of news articles in the American press (and the British also) detailed the lurid adventures of a British salvage ship seeking the *Lusitania* —fantastic stories which all through a long summer kept the English investors who had put up the money keenly on edge for the rich harvest which was soon to be garnered from the deep, ridiculous stories, helping to set the stage for a similar financial raid on gullible Americans. What happened? In spite of all the yarns concerning the fabulous efficiency of the gadgets employed, nothing at all. But by a gratifying coincidence, coming just at the right moment to keep the "investors" properly keyed up to contribute for another campaign, on about the last diving day of that season, the *Lusitania* was "found." But she might just as well have stayed lost, for the British public apparently had enough and has so far refused to put up sufficient funds again to permit the salvagers to go back and retrieve the treasure. Perhaps America will do better in coming to their aid.

Is there any treasure for salvors on the *Lusitania?* There certainly is. But it seems to lie in the pockets of the public, and in selling to the press and to the radio lurid stories of the search. To paraphrase the old prospector's words:

"Thar's gold in them thar syndicated articles!"

But there is none in the *Lusitania*. And meanwhile real divers and salvage masters open their eyes in wonder, for the whole performance smacks of the late P. T. Barnum's circus stunts rather than of salvage.

CHAPTER XIV

A TREASURE legend that in a sense died "aborning" was that of the *Empress of Ireland*, one of the great marine disasters of our day and of all time. Diving, under difficult circumstances, killed that story in its infancy.

On May 29, 1914, the Canadian Pacific liner, the *Empress of Ireland*, cleared its last port, Quebec, and, loaded down with tourists, headed eastward out into the wide St. Lawrence for Great Britain. Some hours later, during the night, she was in collision with the Danish collier *Storstad*, and, with a huge hole torn in her by the *Storstad's* bow, swiftly plunged to the bottom, taking to their deaths 1,024 of her ship's company, mostly sleeping passengers.

The wreck of the *Empress of Ireland* was the immediate scene of diving operations, but diving conditions were difficult. The wreck lay on its side, the depth was over 180 feet, the tidal currents in the St. Lawrence were swift, and the water was extremely cold. Nevertheless, in an attempt to recover about a quarter of a million dollars in silver aboard, the mail she was carrying, and (most of all) the bodies of the thousand passengers whose coffin she was, the Canadian Pacific authorized diving operations, the work being handled by a Canadian salvage company.

The divers had no great difficulty in recovering the

silver and the mails, in spite of the depth and the current, but the remainder of their task was perilous.

In a maze of tangled wreckage, cabin furniture, mattresses and bedding, the divers groped on hands and knees along the bulkheads, worming their way through passages no longer high enough to stand erect in, searching through the cabins for the drowned passengers, whom they sent up by the hundreds, each at the end of a line. Under circumstances as morbid as any ever encountered on the bottom, the divers worked their way through the ship, with men and women, suddenly overtaken by death, staring at them with sightless eyes and agonized faces almost at every turn. To make matters worse, the ship was soon alive with shellfish attracted by the feast, and many a corpse went up so badly eaten as to be forever unrecognizable.

In spite of cold water, in spite of "the bends," the divers kept on through that wreck. Death seized one of them, an unfortunate killed by a "squeeze" when he plunged from the bow of the wreck into the greater depths below, there to be instantly jammed into his own helmet by the sudden increase in water pressure. Still they stuck to their task till the mail was salvaged, the silver recovered and the bodies of the passengers sent ashore for burial—many, never identified, to rest in a group huddled in the shadow of a massive stone cross erected on the south shore of the St. Lawrence opposite the point in the bay where lies the *Empress of Ireland* herself.

Meanwhile a new problem arose to plague the owners of the ship and the underwriters. From the heirs of those lost on the *Empress of Ireland* claims by the score started to pour in on the distraught shipowners for jewels and for money taken aboard the *Empress of Ireland* by the lost passengers and there confided to the purser's safe for security. Buttressed by legal documents and affidavits, the claims poured in, and the amounts claimed rose to disturbing proportions. As in the case of the *Merida* and the *Lusitania*, the *Empress of Ireland* was fast becoming another treasure ship.

Faced by the prospect of an additional unlooked-for loss, the underwriters determined, costly though it was, to recover the purser's safe and return the valuables to the legal heirs, thereby escaping a much heavier expense in settlement. So on top of his other troubles, the salvage master was ordered to retrieve that safe unopened, undamaged, and intact, so that no claimant could possibly say that rascally divers had looted it on the bottom, to bring up the plunder clandestinely.

To perform this herculean task, the salvage master and his divers studied, as she lay at her dock in the St. Lawrence, a sister of the ill-fated *Empress of Ireland*, carefully measuring the safe there, the hatchways going up and all the surrounding corridors. The *Empress of Ireland* lay on her side; it was hopeless to try to get that safe up through the hatches in the superstructure. Up, so far as the force of gravity was concerned, now meant through her side, not through her decks. But they finally puzzled out a way, and back they went to their wreck to carry it out.

Handling a large safe is always an awesome task, as the size of the crowd on any city sidewalk, gaping as the riggers roll it off a truck and hoist it skyward, certifies. In this case, the safe-movers had to get their safe out of a half-capsized ship, over to her side, and then through that and up through a depth of water equal to the height of a 16-story building. Quite a problem.

On their hands and knees on top of that heeled-over safe, the divers in their cumbrous rigs struggled, in the darkness and the cold of the water, in the pursuer's office, fumbling with wrenches over the treasure chest beneath them, feeling out its edges, then with heavy wrenches unbolting the massive fastenings which held it to the bulkhead behind.

That done, cold and frozen, they went up, while a second set of divers with crowbars and sledges smashed away the surrounding furniture and then laboriously heaved the safe over on its side so it might go through the door into the passage above.

Meanwhile, on the thick steel side of the *Empress of Ireland's* hull, a side no longer vertical, but facing the surface, other divers felt carefully over the rivets, portlights and fittings till they located a side cargo port almost vertically above the purser's office. Here they planted a light charge of dynamite and blew open the port, giving them a hatchway down which a diver ran a hawser from the salvage ship.

And now the rigging work really started. Out of the purser's office, forward through the water along the cramped corridor, barely passing through it, went the safe, dragged along by the winch on the salvage vessel in the world above, till it came at last opposite the opened port. There the divers, for the last time, levered

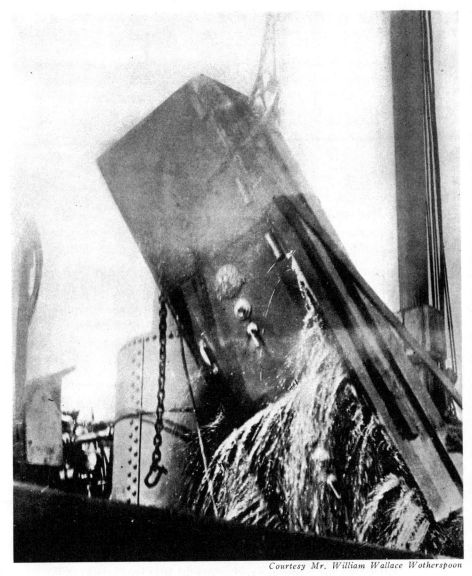

Courtesy Mr. William Wallace Wotherspoon

THE PURSER'S SAFE, SALVAGED FROM THE "EMPRESS OF IRELAND,"
COMES OVER THE SIDE

it about and rerigged it for hoisting, then got safely clear and up. Their measurements had been correct, the safe had moved steadily out without jamming.

From on deck, the salvage master directed the last lift. The steam winch groaned as its pistons labored under the load. Bumping and banging against the bulkheads in the *Empress of Ireland*, the safe careened from side to side in its erratic rise, with the strained hawser quivering as the shocks of striking and bounding clear telegraphed up it. One final bump and the safe came clear of the wreck, turning slowly in the water after that as the winch heaved in and it rose steadily to the surface.

It was a long lift. Every eye on the salvage ship was glued anxiously to the taut hawser. Then a splash, a cloud of spray, and, amid a fountain of water spraying from every seam, the precious safe rose clear of the water. A cheer burst out from the divers, then suddenly faded as they noted that the lifting line had somehow slipped up the safe, was within a few inches of the top! If it slipped a trifle more, the safe and its valuable cargo would slide back into the sea! With bated breath they watched as the salvage master slowed the winch, gently, very gently, swung the boom in, lest he lose his load, and then gasped with relief as the safe which meant so much in desperate labor to them cleared the gunwale and was dropped, secure at last, to the deck. Then indeed they cheered. That safe was priceless!

With water still pouring out of its seams, several officials, representatives of the Canadian Government, immediately stepped forward and put seals upon it.

Carefully then it was transferred to another ship, and, still under heavy guard, transported to Montreal where it was sent for safekeeping to one of the largest banks, completing a salvage exploit of which both divers and salvage master might well be proud.

A few days later, in the presence of the legal representatives of all concerned, in a room so crowded with barristers representing claimants that there was hardly space for movement, the massive safe, risen again from the depths, was opened, by the use of the combination known only to the line's officials. And the gasp that went up from the divers when they nearly lost the safe was as nothing compared to the gasp that went up from the claimants when the contents of that safe were exposed!

Of course the valuables inside were trifling, not five per cent of the huge sum claimed. Papers, yes; some worthless letters of credit, on which recovery had already been had, yes. But real money and sparkling gems? Not very much. It was laughable, for the contents hardly covered the expense of tearing the *Empress of Ireland* open to recover the safe. Right there the treasure legends of the *Empress of Ireland* died in their infancy.

It would seem that pursers safes always are stuffed full of the treasures of Golconda—until they are opened.

CHAPTER XV

THERE really are and have been some rewards for men under the sea. Whenever a real, authenticated, indisputable treasure has gone down, through the centuries, divers and salvagers have gathered there within a reasonable time, to recover all or a good part of the golden hoard. Technical difficulties, physical dangers, the fury of nature, death itself, have never deterred them. The major difficulty in treasure recovery has never been in providing the technical equipment—the supposed marvels of science which cracked inventors and their promoters are continuously dangling before the eyes of the gullible as the new-found key to the lost treasures of the deep to the contrary notwithstanding. The major difficulty has always been in finding an authentic treasure worth working on with the equipment already available. When such a treasure is really lost (in fact, not fable), it is amazing to what lengths men have gone in recovery and what successes they have achieved under the sea.

One of the earliest and queerest adventures in which divers have recovered a real treasure is mingled with the early Colonial history of Maine and Massachusetts. Like many another New England youth of his time, William Phips went early to sea. He had to, for, born in 1651, he was one of a brood of children (all born to the same mother) so large that it was doubtful that all his twenty-five brothers and sisters ever gathered at one time for a family meal.

William Phips as a boy grew up as a shepherd among the Maine hills fringing the Kennebec, then served an apprenticeship as a shipwright and got his first love of the sea in helping to build the wooden ships that were springing up on the stocks in nearly every one of Maine's rugged inlets. Migrating to Boston, a full-fledged shipwright, at the age of twenty-two he first learned to read and write. Soon after he started his career as a fortune seeker by marrying a wealthy widow considerably older than himself. At this event, as usual, there was much shaking of heads and excited gossip among the good Puritans of Boston Town (especially the ladies) over Phips' motives.

Phips concerned himself little over what the neighbors thought. His motives were perfectly clear to him, and without delay he acted on them. Taking his wife's money, he promptly bought himself a small vessel and blossomed out as a ship captain, his carpentering days behind him forever.

Commerce in Boston at that time was principally with the West Indies, lumber going south and rum (or the molasses for the making of it) coming north to godly New England.

On his voyages to Hispaniola and Cuba Phips heard much of wrecked Spanish galleons, in particular of one treasure vessel, flagship of a Spanish vice admiral which, some forty years before, had been driven on the reefs to the southward of the Bahamas, and there lost with many of her crew. This was recent enough for young Phips to be able to verify the fact that the ship had actually once existed, so, bidding good-by to the trade in lumber and rum, he departed for the Bahamas. There he found the facts to be about as stated. The waters around those islands were clear, warm, and shallow. Naked divers working from canoes had over many years recovered here and there silver ingots and pieces of eight, and there were actual bars and coins to prove it.

Phips joined with other adventurers (hardly more than pirates) and went looking for a galleon, but he found nothing. However, having previously been both sober and industrious in managing his wife's estate, he had preserved enough of her fortune to permit him to acquire a few salvaged silver ingots as evidence. With these he set out for England, armed in addition with a chart showing the location of the vice admiral's galleon that he himself proposed to salvage.

Phips, bronzed by the tropic sun and overflowing with tales of Spanish silver, was intriguing enough a personality soon to get the ear of the Duke of Albemarle. This exalted nobleman introduced him to no less a personage than King Charles II himself.

Charles II, king by the grace of the death of Oliver Cromwell, and one of the most dissolute and profligate of English monarchs in all history, was in desperate need (as always) of money, and lent a ready ear to Phips' tales of huge piles of Spanish ingots nestling among the Bahaman coral reefs. Phips could not have found a more willing listener. He needed hardly to sell his story to the King—that impecunious monarch sold himself so thoroughly that he made William Phips captain of a British frigate, the *Rose of Algier* of 20 guns, and promptly sent him forth to retrieve the treasure, armed in addition with a royal warrant bidding all and sundry subjects of His Majesty under penalty of royal displeasure to render Phips what assistance he required in that mission. To Phips' chagrin, however, there was inserted another clause in his commission which griped him severely. Since the *Rose of Algier* was not on a naval mission, even though she was a warship, he was forbidden to require other vessels to strike their colors to him, an honor otherwise due him as commander of a British man-of-war.

With this single drop of gall in his otherwise overflowing cup of cheer, Captain William Phips set out for the Bahamas by way of Boston, and promptly proceeded to ignore that part of his commission which displeased him. Staid Boston, which before had known Phips only as an obscure shipwright and an insignificant merchant captain, awoke one day and rubbed its startled eyes at the sight of its former townsman in command of the *Rose of Algier* firing solid shot promiscuously across the bows of every vessel moving in or out of Boston Harbor, with the peremptory demand that it strike its colors to his flag on pain of being sunk.

If this performance left any of Boston's citizens unaware of William Phips' new glory, they soon enough learned of it otherwise, for Phips' crew, as fine a set of unhanged scoundrels as ever disgraced the sea, quickly

had Boston in an uproar with their carousing, their profanity, and the sight of such sober citizens as ran afoul
them seeking shelter up the lamp-posts to escape their
drunken assaults. Phips, called to account for the conduct of his crew, went his men one better by his own
conduct in court, roundly cursing both magistrate and
governor for daring to intervene. But the governor, having glimpsed Phips' commission, promptly denounced
him and fined him heavily. That Phips and his ruffians
got off only with fines instead of ending in Boston jail
can be laid mainly to a healthy fear of what the dreaded
King Charles might do should his treasure expedition
be thus summarily concluded with its accounts still decidedly in the red. So, in spite of their crimes, the governor released the culprits and permitted Phips to sail
for the Bahamas, while distraught Boston pulled itself
together, held an unofficial day of thanksgiving on his
departure, and prayed earnestly ever to be spared another visit from their former townsman.

Phips himself, in no wise chastened by the governor's
denunciation, sailed on to the southward. Soon he
reached the vicinity of the Bahamas, and the search for
a treasure galleon was begun in earnest.

For weeks, the little *Rose of Algier* cruised around,
but the only wreck Phips came across was one which
had already been picked over rather thoroughly in years
gone by, and which even then had some six other
treasure-hunting vessels hovering near by cleaning up
what little was left. These ships Phips chased off, over-
awing their skippers by his royal commission, backed
up by his own bluster and the guns on the *Rose,* to start

diving himself. But a trifling amount of silver was all
the reward he ever obtained from that wreck, and, soon
abandoning it, he continued his search for the vice admiral's galleon of which he had heard. There among
the Bahaman reefs and shoals to the northward of
Hispaniola, with his assorted collection of cutthroats
eagerly manning the bulwarks and peering downward
into the transparent Caribbean waters, looking for the
hulk of the foundered galleon, Phips and the *Rose of
Algier* searched long and earnestly. But as the weeks
dragged on, supplies and water, never good, became
worse under the tropic sun. With no sign of a treasure
ship to be looted, the crew became first restless and then
threatening, and Phips had good cause soon to concern
himself with matters other than sunken Spanish galleons.

An armed delegation of his crew waited on him on
his own quarterdeck with a very businesslike suggestion. The expedition was out to loot sunken galleons
of their treasure, but, having found none, why waste
further time? Galleons not yet sunk were easier to find.
Not so far to the southward lay the Spanish Main with
galleons occasionally laden with Mexican treasure homeward bound for Spain. Phips had a man-of-war of 20
guns. It was necessary only to forget the King's commission and hoist the black flag; the piratical mob before
him pledged their heads there would soon be treasure
in their hold and wrecks enough littering the bottom
of the Caribbean to satisfy all hands.

Such a villainous proposal may sound odd to us in
1939; in 1685 to Phips' crew it seemed quite reasonable,
especially in view of the fact that only fourteen years

before the redoubtable Henry Morgan had looted Panama City under not very different circumstances, and,
having returned to England with gold enough, had in
consequence become Sir Henry and was at that very
moment acting governor of Jamaica.

Still, to the unmitigated surprise of his seamen, un-
impressed by the suggestion and unintimidated by the
bared cutlasses and knives confronting him, Phips, unarmed and barehanded, suddenly rushed the men before
him, knocking the leaders into the scuppers with broken
heads and causing those behind to flee as from a mad-
dened bull.

Thenceforward, no one dared any further suggestions
to their brawny captain, and Phips, unmolested, kept
on searching through the shoals as directed by the
King's commission. But, commission or no commission,
he had ample reason now to believe that, unless he
quickly discovered some treasure, it was most likely
that his bones would shortly be whitening amongst the
coral so plentifully in view everywhere he looked for
silver.

However, no treasure ship loomed up through the
sparkling depths to save the situation, and his unruly
crew grew daily more insubordinate. To add to Phips'
troubles, the bottom of the *Rose* became so fouled with
marine growths from her long sojourn among the tropical islands that she could scarce sail, and Phips was reluctantly forced to call a halt in the search while he
beached his ship on a small wooded island to careen
her and scrape her bottom.

With most of the crew over the side for the scraping, well away from their captain's watchful eye, the more violent among them seized their opportunity. They gathered all those who were ashore among the palms a little back from the beach, during a halt in the scraping for rest, and plans for seizing the ship were quickly laid and enthusiastically assented to by all hands present. When they went back, they were, instead of scraping, to rush the vessel via the narrow gangway that spanned the strip of water between ship and beach, toss overboard the captain and the eight men still aboard with him, to sink or swim as best they might, and sail away with the well-armed *Rose* for their piratical cruise.

But one thing more they needed. Little as they liked him, they had to have the ship's carpenter with them in their future activities, and, as there was a fair chance he might be killed battling at the captain's side to save the ship, they determined to get him ashore first, where he could be held safe. On pretext of some scrapers needing repairs, he was quickly lured down the gangway and into the woods to inspect them, only to find himself in the midst of dozens of flashing knives, with the alternative of joining the mutiny immediately or having his throat cut forthwith.

Knowing his single-minded shipmates only too well, the unfortunate carpenter joined without hesitation, whole-heartedly commending the mutineers for giving him the opportunity to join such a promising enterprise, but pointing out to them that, as he had brought none of his tools ashore for the supposed repairs, if he did not quickly return for them, the captain (hav-

ing already reason enough for alarm) might well become suspicious of the delay and hoist in the gangway,
meanwhile putting the ship in a state of defense which
might completely frustrate their designs. If, on the
other hand, he went back for his tools, the subsequent
delay in their recommencing scraping the ship, the
while they concluded all preparations for the attack,
would naturally be attributed by the unsuspecting skipper to the fact that he was still busy repairing the scraping equipment.

The mutineers considered this suggestion dubiously,
properly enough fearing treachery, but the carpenter's
logic was sound and his enthusiastic praise of their plans
was more than convincing. So, sending two seamen with
him, ostensibly to carry the tools but actually to prevent him from speaking to anyone aboard, they permitted his return.

Out from the palms, across the sandy beach, and
up the gangway of the nearly deserted *Rose* went the
carpenter followed by his two warders, their knives
conveniently loosed in their sheaths. The captain was
aft in his cabin and his other petty officers were below
decks, so no one aboard was anywhere near as, osten-
tatiously, the carpenter unlocked his chest and began
rummaging through his tools, occasionally lifting one
out and handing it to one or the other of his two guards
until their hands were full. Then suddenly, with an
agonized expression on his seamed countenance, the
carpenter doubled up over his opened tool chest, exclaimed he was taken with such a cramp in his bowels
that he was like to die unless he had medicine instantly,

and, before either of the two slow-witted sailors with his arms loaded with tools could drop them and reach for his sheath knife to restrain him, he was off in the direction of the cabin, where, as all hands well knew, the medicine chest was stowed.

In on his captain burst the carpenter and in a few brief words disclosed the danger, the while he kept pointing to his supposedly anguished intestines. Phips rose to the occasion. Undismayed, and in full sight through the opened cabin door of the two astonished seamen who a little distance off were still clutching their armfuls of tools, uncertain whether to stay or to run, the captain went calmly to the medicine chest, reached into it for a dram of some innocuous medicine, and poured it out for the carpenter, meanwhile cautioning him in a quiet tone to go back as if nothing had happened, leaving all to the captain.

Wiping his mouth and making a wry face over his medicine, the carpenter mumbled out his thanks and, still groaning but now not so much, rejoined his guards and shortly followed them down the gangway. His escorts were so completely taken in by his manner and by his unexpected return that neither one in the least suspected or reported anything to the mutineers awaiting them in the woods.

Aboard ship Phips acted swiftly. His guns (as always, because of fear of Spanish or piratical attacks) were loaded, but he made ready the gunports on his careened ship for instant casting loose, laid out extra charges, lighted his slow matches, and stationed a few men to hoist in the gangway at the signal. Then he waited.

Perhaps an hour later, his crew came straggling out of the woods, gathered into a compact group, and then started to cross the sandy beach toward the ship. Phips made no move till all were completely clear of the shelter of the trees and within easy range, when, at his signal, up went the gangway, open flew the gunports, and the dumbfounded mutineers discovered themselves looking across an unfordable strip of water into the menacing muzzles of their own guns, with matches glowing above the touchholes, ready to fire!

Huddled together like sheep before the butcher, the cowering mutineers halted in their tracks. They could neither attack nor escape. Taunting their helplessness, Phips on pain of death bade them stay where they were, the while with his few men he prepared to make sail, leaving his mutinous crew without food or water to starve or die of thirst on that barren islet.

That prospect quickly brought them round, as Phips well knew it would. On their knees, the panic-stricken mutineers begged not to be left marooned, promising piteously implicit obedience to Phips' every order, and pleading in extenuation that only his obstinate opposition to turning pirate had driven them to mutiny.

Phips, with only eight men aboard to help him work his frigate back to an English port, was in almost as bad a spot as his mutinous crew and was only too anxious to compromise with them, but he carried off all with a bold front till the men before him seemed reduced to a properly importunate state. Then, finally, with seeming reluctance he relented, lowered the gangway, and, in twos and threes, let the men come aboard.

Ironing the ringleaders, and depriving the more dangerous of their followers of weapons, while arming those in whom he still had a little faith, he promptly made sail for Jamaica. There he completely discharged all who had taken part in the mutiny, shipped a new but smaller crew, and once more, able to breathe a little more freely, headed the little *Rose of Algier* back toward the coral banks north of Hispaniola.

For many more weary months, Phips continued his search, but nothing did he ever find. Finally, with supplies exhausted and his crew worn with sickness, he gave up and returned to England with the frigate *Rose*.

Three years had gone by since he sailed. Charles II had died, James II sat on the throne. In anger at the lack of treasure, James canceled Phips' commission, took his ship away, and levied a charge of £500 against him for repairs to the *Rose,* leaving him with nothing to show for his years of searching except his chart, still decorated with a cross showing the location of his unfound galleon.

James II would, however, pay no further attention either to Phips' charts or to his stories, so Phips was forced to look for other sponsors. There followed trying months for him in London while he sought backers, a period in which among other trials he landed in jail as a result of claims against him. But Phips had firm faith in the existence of his galleon, and that faith carried him through a difficult time in London. After long months of effort, Phips finally persuaded the gentleman to whom first he had brought his story years before, the Duke of Albemarle (and some few of his friends), to finance another ship, the *James and Mary*, in which, accompanied by a small tender of 40 tons, he set out again in 1686, the gentlemen adventurers having subscribed £3,200 for expenses.

A little wiser now than formerly, before starting his search, Phips built himself a large Indian canoe, which, much shallower in draft than either his new ship or its tender, permitted the search to extend among reefs where he had never dared to take the *Rose*, and where neither of his present vessels could safely penetrate.

Putting in again at Porto Plata in Hispaniola, Phips managed gradually to worm from an old Spaniard, who originally had given him some information about the vice admiral's galleon, further details of the wreck as that aged colonist had obtained them over forty years before from some few of the survivors who had escaped on a raft. Learning enough to localize the reef on which the vice admiral had come to grief as being the Ambrox or Ambrosia Reef, Phips sent his tender and the small boat off to search that reef while he did a little trading in Porto Plata and replenished supplies.

Phips' expedition was for its day well-equipped for diving, being provided with all the latest marvels of science. He had with him a "diving tub," an early form of diving bell. This particular device, one which probably he had himself devised, was nothing more than an inverted shell or tub, weighted to hold it down, beneath which a man might keep his head and shoulders in such air as it contained while under water, occasionally stepping out from under it for a minute's work.

For divers, Phips had four men with some knowledge of the diver's art as it had been practiced up to that time for several thousands of years. Diving was simple in those days. The diver, whose principal requisite was a good chest expansion, started by taking a huge breath to fill his lungs and then went overboard headfirst, naked or practically so, clinging with one hand to a weight to carry him down, and with a line round his waist for his attendant to bring him up afterward. He might go, if skilled, as far as 75 feet down, the increasing water pressure squeezing in his ribs as he continued his descent. The limit of depth was set for any man by the point at which further decrease in lung volume would result in his ribs cracking instead of bending in to match his much-reduced lung capacity. At around 75 feet, the total pressure on the diver becomes approximately 3 atmospheres, and consequently the air in his lungs occupies one third the volume it had when he left the surface. Correspondingly his ribs must flex in to reduce his chest capacity similarly. If they are unable to bend in sufficiently, either water enters the lungs, to compensate, and drowns the diver, or his ribs break, puncturing his lungs, with death then ensuing from internal hemorrhage.

In the normal case, with the naked diver within the range of depth to which his chest can accommodate itself, the diver arrives on the bottom with a much-shrunken pair of lungs, and with his buoyancy much reduced thereby from what it was when he left the surface. He is now sufficiently heavier than the surrounding water to stay submerged without effort and even without his weight.

And now the diver must become exceedingly active, for his total stay beneath the surface can hardly exceed 1½ minutes, that being about as long as the average naked diver can hold his breath. With his vision considerably blurred by water in direct contact with his eyeballs, the diver does what he can during his brief stay, usually letting his now useless weight go while he works. When he can no longer hold his breath, he signals his tender by jerking on his line, and is immediately hauled up, speed in ascent and descent being vital to give to the work on the bottom as many as possible of the precious 60 or 90 seconds available to spend beneath the sea, and as few as possible to the unavoidable travel up and down.

The naked diver need never worry about decompression or "the bends." It is physically impossible for him to remain below long enough for his blood, even under the increased pressure, to absorb any harmful amount of nitrogen. However, he has enough other worries, being unduly exposed to hemorrhages from the continued strain of holding his breath, and the likelihood of becoming unconscious under water from excessive carbon dioxide which gathers in his lungs while his breathing is suspended.

From the salvager's viewpoint, the principal shortcoming of the naked diver is his inability to stay down long enough to accomplish much work per dive and consequently to penetrate very far into a wreck. However, in Phips' day, ships were simple structures and their holds were not much divided, making access relatively easy as compared with modern vessels. So Phips, with his "diving tub" and his divers, no doubt felt himself well-equipped for his task and as up to the minute scientifically as human ingenuity would ever get.

More weeks went by with no report from his absent tender, and Phips learned what many another salvager has since discovered—that treasure recovery depends only to a trifling degree on scientific marvels for diving; the real problem is in locating an authentic treasure ship. Of the authenticity of his wreck, Phips himself had no doubts, but when in God's name was he ever going to find her?

Meanwhile his two small auxiliaries cruised over the Ambrox Reef, carefully scanning every yard of it. It was an odd formation of coral, with a steep face against which a vessel might strike only to sink in deep water alongside, and looked not very promising for the divers should they find anything there. The men faithfully searched it yard by yard, nevertheless. Finally one of the seamen, noting through the clear water what appeared to him as a peculiar marine formation in the coral far below their keel, ordered a diver over to examine it.

Weight in hand, down plunged the diver headfirst through some 6 fathoms of water to the bottom, only to come up excitedly almost immediately with the strange news that he had sighted down there, coated with coral, what he claimed were a number of guns!

Incredulously the other seven men in the boat eyed him, for, from above, the reef there looked no different from hundreds and hundreds of other coral patches they had previously sighted; but the diver was certain they were over a ship with its open hold directly beneath them. As soon as he had recovered his breath, he dived again to prove it.

And he did, for when he signaled to be hauled up, he was clutching in his arms a coral-crusted pig of silver!

Deliriously his shipmates fingered that dripping silver bar. It was silver beyond a doubt, and they had found their galleon! Quickly marking the spot in the reef with a buoy, they feverishly made sail for Hispaniola to rejoin Phips.

Phips, loading stores in Porto Plata, displayed little enthusiasm when his gesticulating seamen rushed over the side of the *James and Mary* to inform him they had found the galleon, for he himself had found several worthless or already looted hulks in his long years of search; but when the silver ingot was proudly produced, his excitement equaled theirs and, as rapidly as was possible, the *James and Mary* was got underway for Ambrosia Reef, since renamed the Silver Bank.

With the ship swinging at anchor as near as he dared take her to the dangerous reef, Phips in the small boat accompanied his men to the designated location over the reef. Down plunged the first diver, his naked body waveringly visible to Phips' feverish eyes as he scrambled 6 fathoms below over what seemed only rough coral. But up with their diver came another silver ingot, and fortune had smiled at last on William Phips!

They were indeed over the waist or midship hold

of their long-sought treasure ship, and down there in from 6 to 8 fathoms of water lay piled up heaps of ingots. One after the other, Phips' four divers went overboard, each time to come up clutching another silver bar. When night fell on their first full day of diving and the exhausted men had finally to quit, the small boat returning to the *James and Mary* was nearly awash under her load of bullion.

For a month Phips and his crew worked in relays over the hulk, hauling up silver, limited only in their speed of recovery by the negligible length of each dive. Phips tried to expedite matters by having his men stay down longer, working with their heads beneath his "diving tub," but so uneven was the footing which the hulk presented inside that this was a failure. Like many another scientific marvel of later years, it had quickly to be abandoned, leaving the recovery wholly to the divers and to such improvised tongs and grapnels as could be employed from the surface while the exhausted divers were catching their breath and stanching their periodic hemorrhages.

In the midst of this feverish activity on the small boat, the *James and Mary*, guns always ready to repel either Spaniards or pirates, lay close to the wreck, the pile of bullion in her hold steadily increasing. Phips, with provisions and water low but afraid to leave— in deadly fear that wind of his success would thus get out and soon bring down on him a Spanish squadron or, what was worse, a swarm of buccaneers—was in great haste to clear the wreck and depart; so not for a single day when the weather permitted did the treasure lifting stop to give his strained and bleeding divers a real chance to rest and recuperate.

Other things beside silver bars came up. The divers soon noted in the hold what seemed to be large lumps of coral, and sent one of these up. When it was smashed, it was found to be a mass of silver coins encased in a coral-crusted canvas bag. After that, coral lumps and silver pigs came up indiscriminately, till in weight the *James and Mary* had aboard as much coined silver as she had bars.

For over a month monotonously the divers toiled. One after another in endless succession their naked bodies flashed in the sun, plunged 40 feet downward through the warm sea, groped a moment or two inside the whitened hulk of that galleon, and, clutching whatever came to hand that was sufficiently heavy to be silver, were hauled to the surface, there to sit gasping for breath again in the sun while another coral-coated silver pig or bag of coin went to join the ever-growing treasure in the *James and Mary*.

Phips had long since quit worrying over his divers and their ability to cope with the salvage problem, but as the wealth in recovered silver started to pile up, he began to worry about his crew. It took little perception to read trouble brewing as their greedy eyes daily scanned the heap of silver tumbling aboard from the wreck. For in all that treasure, they were not to share, having been signed on by his parsimonious group of

London gentlemen adventurers for wages only, and very skimpy wages at that. That the rough seamen under him might be waiting only till the last bar of silver was recovered, to seize the ship and reimburse themselves as they chose, was more than probable. And in the murmurings and grumblings of his crew against their scanty rations and grueling labor, Phips read impending disaster. With recollections still vivid in his mind of his previous experiences with disgruntled seamen, Phips had little wish to face another mutiny in which he might not come off so well, and he took steps to head it off.

Gathering his crew together, he promised them over and above their wages a bonus from the treasure when it should arrive in London. And he pledged his men that, should the proprietors refuse to pay it, he himself would disburse the reward from his own share, so that every man might be assured of getting what was offered him as his portion of that heap of silver in the hold.

Probably the timeliness and the generosity of Phips' personal pledge had more to do with making his offer acceptable to men who after all were little removed from being pirates, than the sums offered, but at any rate no more of mutiny was heard again; and, when the main hold of the sunken galleon was finally cleared of treasure, Phips sailed away from the Ambrosia Reef unmolested by fears of his crew.

But his troubles were not over. He dared not enter any port to provision for his Atlantic passage. In any Spanish port his treasure would certainly be seized as belonging to the King of Spain; and he too well knew the rapacity of British colonial governors to trust himself to any of them. There was no way out but to go home with what scant rations he had aboard; and that he did successfully, arriving in the Thames off London in September, 1687. In his hold were 34 tons of silver bars and coin, a few ingots of gold, some pearls.

The sensation created in England by the arrival in London of the *James and Mary* with that cargo was immense. Phips became instantly a popular hero; the Duke of Albemarle and his gentlemen adventurers, wealthy beyond their fondest dreams. So overwhelmed were they by what Phips had recovered that without demur they paid the crew the bonus Phips had promised them.

The treasure of the vice admiral's galleon was then valued at around £200,000—in present day equivalents, well over three times that, or about $3,000,000. Phips' share, one sixteenth only, was for that day a very considerable fortune. In addition, so impressed was James II at the persistence and courage of his loyal subject, William Phips, that out of hand he knighted him, and soon after made him royal governor of Massachusetts. The King, by the way, who was out of pocket nothing in this expedition, nevertheless collected a royalty of one tenth, or £20,000, on the undertaking, which may have had something to do with his seeing qualities in Phips which a few years before, when he took away from him the frigate *Rose*, he was wholly unable to discern.

And so Sir William Phips, to the great astonishment of his fellow citizens of Boston who had last seen him barely escaping the stocks, came back to Massachusetts in much state, by virtue of his own good qualities and the diving abilities of a few naked seamen.

CHAPTER XVI

AMONG achievements of men under the sea, that on the *Egypt* will always shine as of the first magnitude. Death and disaster, the weary waiting that wears down the heart, legal entanglements, the terrors of the deep sea and glittering bars of gold are woven into that story as in none other.

For me the story began in New York and ended in Rome, with seven years between. During that time death had taken heavy toll of many of the actors in the drama, and finally and ironically had laid hands on the man (himself an expert in deep water diving) who first called my attention to the *Egypt*, the deepest salvage job ever undertaken.

I found on my desk one day in early June, 1929, a foreign letter. I opened it curious, read:

Paris, June 1, 1929.

SIR:

I am trying, by this letter, to introduce myself to you, as an Italian Navy Lieutenant, in charge of some experiences for Submarine Salvage and Rescue Works.

I have just now read your book "On the Bottom" and some articles of yours in the *Saturday Evening Post*, and I have very much learned from your vast experience.

I should be most grateful if you would let me know what books have been written by you and other experts, by reading which I may complete my own experience.

I am going now to see the rescue work on the British Steamer *Egypt* which lies at the bottom in a deep of 360 feet, off Brest (France). Let me know if there is some interesting for you in that job.

Sending you my best regards and thanks, believe me,
Yours sincerely,
ALBERTO CUNIBERTI, LIEUT. I.R.N.
c/o Italian Naval Attaché
Italian Royal Embassy
Paris.

Commander Edward Ellsberg,
Bureau of Navigation,
Navy Dept., Washington, D. C., U.S.A.

Something interesting for me in a salvage job at 360 feet? There certainly was, though I was incredulous. I smiled a little at Cuniberti's English, but, after reflecting that I couldn't do a tenth as well in his Italian, I hastened to write him. And so commenced a correspondence lasting some years, which kept deep diving, the *Egypt* and what went on regarding her before my eyes. Cuniberti, I learned by checking elsewhere, was the leading expert in submarine rescue work and deep diving in Italy, and well worth knowing.

The S.S. *Egypt*, as I soon found out, had been a P. and O. liner of moderate size, some 7,941 tons, plying between London and India. She sailed on her last voyage for Bombay in May, 1922, carrying a passenger list of 44, a crew of 291 (mainly Lascars), and a cargo of bullion in her strong room consisting of 1,089 bars of gold, 37 boxes of gold coin, and 1,229 ingots of silver, amounting altogether to about 5½ tons of gold and 43 tons of silver, the whole valued and insured (mainly by Lloyd's) for a total of £1,058,879, or approximately $5,000,000.

On the evening of May 20, 1922, while traveling slowly south-southwest in a dense fog over a perfectly calm sea some 20 miles southwest of the island of Ushant, the *Egypt* heard on her port side a steamer whistle. This was soon followed by the sudden appearance close aboard of the bow of another ship, which, before the *Egypt* could do any maneuvering, struck her amidships between her two stacks, tearing wide open her side, and then faded from view again into the fog.

Immediately the stricken *Egypt* heeled to port and began to sink on her side. While her panic-stricken Lascar crew desperately struggled to lower the boats from her rapidly listing decks, a continuous stream of SOS calls went out from the *Egypt's* radio room, followed soon by her position as hastily computed by her captain in the confusion as his sinking ship settled under him:

"SOS–SOS–SOS–position 48° 10' North, 5° 30' West, *Egypt*."

In twenty minutes after the collision, the *Egypt* was gone, taking with her 17 passengers and 71 of her crew. Very few boats were ever launched. Only the fact that the ship which struck her, the French freighter *Seine*, succeeded in finding her again in the fog before she vanished, and fished many of her floundering crew from the water, prevented a much larger loss of life.

Twenty wild minutes, and the *Seine* floated alone in the fog and the night. The *Egypt*, her radio calls suddenly stilled, lay with her five millions of dollars in gold and silver at the bottom of the sea in 66 fathoms of water, a depth far beyond any to which men had ever gone before.

Lloyd's paid the losses. To the original owners of ship and bullion, the *Egypt* soon became only a memory. But to Lloyd's, legal owners now of that bullion, she represented their sole hope of recouping part, at least, of a very heavy loss, and it was not long before they had entered into a contract for salvage, undeterred by the fact that the wreck lay 100 feet deeper in the sea than men had ever before plumbed, and at a depth twice as great as any in which divers had previously done any real work.

Compared to the problem of the *Egypt*, salvaging treasure from the *Merida* or the *Lusitania* would have been child's play; but where the *Merida* and the *Lusitania* were immediately abandoned as worthless by the underwriters, the seemingly impossible task of recovering the bullion in the *Egypt* was in all seriousness promptly taken in hand—no real treasure of $5,000,000 is relinquished by its owners without a fight for recovery.

Two problems were involved, one to find the wreck, the second to remove the treasure from her. Oddly enough perhaps to the uninitiated, it was the first problem only which really seriously worried the ultimate salvagers; the recovery they felt sure of accomplishing if they could ever locate the wreck.

In 1923, work started, with a Swedish company engaged to locate the wreck, while two British engineers
designed the diving apparatus for recovery. Further
search was conducted in 1926 by a French contractor,
but by 1928, in spite of an active dragging campaign,
especially in the earlier stages, the wreck had not been
found, and there was little need for diving gear of any
kind. Then matters took a different turn.

In Italy, the Società Ricuperi Marittimi (The Society
for Maritime Recovery), Sorima for short, had demonstrated an exceptional ability in deep water work.
Formed, financed, and directed by Commendatore
Giovanni Quaglia to salvage wrecks in Italian waters,
Sorima, under his direction and using the Neufeldt &
Kuhnke metal diving rig, had successfully salvaged from
the wreck of the English steamer *Washington*, torpedoed near Genoa, several thousands of tons of cargo
in the form of steel and copper bars, locomotive boilers,
and railway trucks. This venture, carried out at a depth
of about 300 feet, was especially profitable in Italy,
where raw materials are expensive and labor is cheap,
but it disclosed above all an exceptional executive ability in the man at the company's head to direct practical
work in deep water. In the light of Sorima's demonstrated strength, the British engineers previously engaged recommended to Lloyd's that the contract be
turned over to Sorima. In August, 1928, this was done.
The terms of the contract were not unusual—Sorima
was to stand all the expense of salvage and was to receive in return 50% of what it recovered.

Sorima banked heavily for success in its diving operations on the Neufeldt & Kuhnke metal diving armor, illustrated in the photograph. In the state of the diving art in 1928, it was impossible for men using the conventional flexible diving rig to go much below 300 feet, or to do any appreciable work even at that depth. The record for depth was 306 feet, reached by three men in the salvage operations on the *F-4* off Honolulu. At that depth, owing to the tremendous pressure, a dive was limited to 20 minutes on the bottom, and no really effective work was asked of or accomplished by the divers. They acted merely as observers to check the position of the steel hawsers swept under the bow and the stern of the submarine by means of which the actual task of dragging the submarine (280 tons weight) into shallow water was accomplished by Lieutenant Furer, salvage officer.

Armored diving rigs had for long endeavored to eliminate the dangers of pressure—"the bends," the effects of highly compressed air on the diver's mind, the long rise necessary for decompression—shielding the diver completely inside a metal suit strong enough to withstand the pressure of the sea at its designed working depth, leaving the man inside to breathe air under normal pressure. Under such conditions, regardless of the length of dive or of its depth, a man might be hauled to the surface immediately when he was ready to rise, as fast as a winch could haul him up, then to emerge from his shell no worse off than if he had been down inside a submarine, for instance.

But while building a shell strong enough to resist the sea is within the capacity of any capable engineer, there were new problems involved in making a practicable rig, which are (as events ultimately proved) still insuperable.

First, no part whatever of the diver's body can protrude from the suit. Here enters the first inescapable defect of the armored rig, its inability to allow a man to use his hands. For if the hands project from the armored suit, they are then exposed to the sea pressure outside, which for a depth of 400 feet, amounts to putting a load of 5 tons on each hand. With no balancing internal compressed air pressure to be transmitted through the lungs to the blood in the hands to offset this sea pressure, the hands will promptly be "squeezed" into jelly, and will as a practical matter promptly be rammed back through the joints into the suit, to be followed instantly by a torrent of water which will drown the diver.

So it is an inseparable feature of any armored rig that a diver using it is deprived immediately of the major feature which has enabled man to rise above the beasts—his hands. All he can do is to substitute for them outside the suit mechanical claws, hooks, or similar makeshifts, to be manipulated as best he may by his hands inside. The first result then of entering an armored rig is that the ability to do any useful work is reduced to a trifling percentage of normal.

A second drawback is lack of flexibility and freedom of movement. In the rubber diving dress, a man can walk, stoop over, lie down, twist himself about, and maneuver arms and legs with nearly as much freedom as in his street clothes, restricted in his ability only by the size of his bulky helmet and the need of keeping his lead weights and his air hose clear of entanglements. All this flexibility is lacking in any rigid dress. To withstand the enormous pressure outside, with no balancing pressure inside, the shell must be made of cylindrical and spherical shapes, with as few joints as possible, for every joint is a source of dangerous leakage. Here comes in another difficulty. Regardless of mechanical ingenuity in designing the joints, as the depth increases the external pressure forces the two faces of the joint more firmly together, increasing the friction of movement and making it harder and harder for the man inside to move either an arm or a leg.

Inspection of the illustration indicates the slight possibilities of ordinary locomotion or of arm movements. What no picture can show is the fatiguing effort necessary to secure even moderate movement erect in this rig in deep water. It requires a physical giant inside to be able to get about at all, and then motion is possible only to a slight degree as compared with ordinary bodily freedom.

Tightness of joints Neufeldt & Kuhnke secured by an ingenious rubber sleeve sliding over a spherical joint, with the pressure tending to hold the rubber edge firmly against the metal, and always in contact with it.

To permit long-continued breathing inside the Neufeldt & Kuhnke rig, the diver wears a breathing mask with a soda lime cartridge to eliminate the carbon dioxide formed by respiration, while fresh oxygen can be supplied as needed from a small cylinder carried in the rig.

Courtesy Lieutenant Cuniberti, I.R.N.

LIEUTENANT ALBERTO CUNIBERTI, I.R.N., WITH THE NEUFELDT &
KUHNKE METAL DIVING ARMOR

Still, in spite of all its limitations, the armored rig, well made as was the Neufeldt & Kuhnke suit, had some value. It could go deeper than a man might go otherwise. Inside it, a man could at least see and act as an observer. If he could be lucky enough to land on a level surface, he could shuffle about a bit, always erect, of course; and, if he were really expert, he could even, with his protruding claws, pass a line about wreckage and perhaps tie simple knots. But as for the diver himself doing any real work from it, that was hopeless. A diver in a flexible dress, within the range to which he could descend, could get into compartments in a wreck and perform more work there in 30 minutes than a man in an armored suit could hope to accomplish in the same spot in weeks, perhaps even in months.

Quaglia, director of Sorima, relied, for success on the *Egypt*, first, on the Neufeldt & Kuhnke diving rig, secondly, on himself and his men. In the end, long before he finished with the *Egypt*, he found his faith in armored diving dresses unwarranted and quit using them; but his faith in himself and his men never wavered, and that finally carried him through, in spite of death and long-drawn-out discouragement.

So in early June, 1929, Quaglia set out to recover the *Egypt's* treasure with his metal diving rig, and his well-tried group of divers and seamen aboard his salvage ship, the *Artiglio*. He carried, as an interested observer for the Italian Navy, its diving expert, Lieutenant Alberto Cuniberti.

The first step, naturally, was to locate the wreck. Where was she? The *Egypt's* captain, as his ship was

sinking under him, had sent out his position as latitude
48° 10′ North, longitude 5° 30′ West. He might ordinarily have been expected to know just where his ship
was, but for some time before the accident he had been
traveling in a fog at reduced speed in an ocean area
swept by strong currents, and his computed dead reck-
oning position might easily be erroneous; besides on a
sinking ship in the haste inevitable in running his last
known position on the chart up to his position at the
instant of collision, an added error might easily have
crept in. What made all this seem more probable was
the fact that two different radio compass stations in
France, as the *Egypt's* frantic SOS calls filled the air,
took directional bearings on her from those signals.
These two stations, one on the island of Ushant, the
other on Pointe du Raz on the mainland south of Brest,
were nearly at right angles with respect to the wreck
and excellently located to get good cross bearings on
her. Their radio compass bearings, when plotted on a
chart, placed her in latitude 48° 6′ North, longitude
5° 29′ West, roughly one mile east and four miles south
of where the Egypt's captain had reported himself.
Further to complicate the matter, a British torpedo boat
destroyer had picked up, a few hours after the wreck,
a mail sack belonging to the *Egypt,* in latitude 48° 14′
North, longitude 5° 30′ 30″ West, some four miles
north and half a mile west of where the *Egypt* had reported herself, and it was possible that beneath that
floating mail bag was the hulk of the *Egypt.*

So, to start with, Quaglia faced the puzzle of where
to begin searching for the *Egypt.* No two positions

agreed within four miles as to where that ship had gone down. In whose position should he have most faith —the *Egypt's* captain, the French radio compass stations, or the British destroyer?

He had one independent report to guide him in his choice. Hedbäck, captain of the Swedish ship which had searched for the wreck in 1923, reported that practically coincident with the position established by the radio compass bearings his drags had hooked an obstruction he took for the *Egypt*, though he had then aboard no diving means of verifying his strike. That the *Egypt* was there, he, Hedbäck, was willing to stake all on.

So in that location, called thereafter Hedbäck's Point, in lat. 48° 6′ N., long. 5° 29′ W., coinciding with the radio compass location, the search started. Quaglia had two ships, his main diving vessel, the *Artiglio*, and a somewhat similar vessel, the *Rostro*. Between the sterns of these two ships as trawlers, he paid out a steel wire cable somewhat over a mile long, held down near both ends by weights. The idea, of course, was that the wire, dragging along the smooth ocean floor, would catch under or on the wreck, thereby indicating its location.

Sweeping started. One end of the wire cable was passed from the *Artiglio* to the *Rostro*, and the two vessels slowly diverged as the cable was paid out, till, over a mile apart with all the wire paid out, they headed on parallel courses at slow speed with the lengthy wire dragging along the bottom between them. It was not long before the sweep made a hard strike, and excitement rose to a high pitch. Had they caught the *Egypt* so soon?

But the sweep slipped free again, and they decided not. Soon the sweep was catching frequently and just as constantly slipping free, which indicated that the bottom, instead of being flat, was strewn with pinnacle rocks, perhaps 30 or 40 feet high, a complication not foreseen on a bottom which the charts showed as practically a level plain some 60 to 70 fathoms deep. A doubt began to grow in the value of Hedbäck's original find in that location. If they had caught not once but many times on obstructions which diving inspections showed to be only rocks, it was highly probable that what Hedbäck had caught, but never seen, was simply another rock.

Weary weeks dragged along while the Italians swept the bottom of the sea, hopefully dropping a man over the side cased in the armored diving suit whenever the sweep caught, cursing fluently each time he reported only another rock. They buoyed off an area 6 miles wide by 10 miles long, with its axis south-southwest, including all the reported positions of the wreck, and set about sweeping that area of 60 square miles systematically. The task was enormous, the difficulty beyond belief. There was no certainty that the sweep might not slide right over the wreck without catching. Stranger things had happened in sweeping. The British, some years before, searching one of their bays for a submarine mysteriously lost, swept time after time the whole bay, only finally to locate their submarine sticking up like a lighthouse almost vertically from the mud in a spot that their sweeps had gone over many times.

Difficulties multiplied. Bad weather and strong currents made it impossible to keep location; the constant rubbing of the sweep on the rocks frayed and broke it, and more time was spent repairing sweep wires than was spent in sweeping. The tempers of the men grew short, discouragement spread about, all hands were ready to clutch at any straw that might help to find the *Egypt*. The wreck was an 8,000-ton mass of steel only 400 feet beneath the surface. Weren't there other means, beside sweeping, magnetic or otherwise, that might indicate where she lay?

There were indeed, and in desperation the searchers for the *Egypt* sought their help. Who in the *Artiglio's* crew had not heard of water and minerals being located beneath the earth's surface by certain psychic persons with divining rods? An idle superstition, perhaps, but the searchers on the *Artiglio* clutched at that straw. There might be something in it. Forked sticks of hazel have a long and a strange record going back to antiquity; even undoubted scientists have sought to explain their somewhat mysterious accomplishments. In Italy especially, their history is ancient, mentioned by both Cicero and Tacitus. If ever the descendants of those old Romans needed help in locating something, the Italians looking for the *Egypt* were those people. They sent back to Italy for an exceptional practitioner with the divining rod, a friar, Padre Innocente da Piovera. Lieutenant Cuniberti speaks very highly of Father Innocent whose incongruous figure he guided about the *Artiglio* while that monk "felt" over the ocean with his forked stick. Of course neither Father Innocent nor his divining rod was of the slightest benefit in discovering the *Egypt*, and back he went to Piovera.

The black arts having ignominiously failed, modern science had a trial. An English engineer who had done considerable work with electro-magnetic apparatus for locating subaqueous masses of metal, such as wrecks and enemy submarines, was next given an invitation to find the *Egypt* with his device. This engineer and his electrical gear were far more impressive than simple Father Innocent and his hazel twigs, but, whatever the reasons, the results were no different; neither ancient magic nor modern electricity could point its finger at the spot where lay the elusive *Egypt*.

So there was nothing left for it but to keep on sweeping by hand, so to speak, and the summer wore away with the *Artiglio* and the *Rostro* still wearily dragging a wire along the ocean floor. Finally the autumn storms came to drive them off; sweeping became impossible in the turbulent seas and they had to quit. The fall of 1929 found the *Artiglio* driven into port with the *Egypt* still unfound.

Sick at heart over their difficulties with the unexpected rocks, Quaglia and his men spent the winter designing and testing a new type of drag to minimize their troubles. They had to have something that would go high enough above the bottom to miss catching on the innumerable rocks, but still low enough in the water to be reasonably sure of catching a ship's hull. They produced a new sweep, made as before of a wire a

mile long, but supported at hundred-yard intervals by
buoys, so that when extended the sweep wire no longer
dragged along the bottom; instead, it floated 25 feet
above it from a string of 18 buoys with a 4-ton weight
at each end of the wire to make sure it stayed down
where it belonged.

With this new sweep, as soon as the weather permitted in 1930, sweeping recommenced. The 60-square-
mile area to be searched was carefully buoyed off again;
the sweepers were rigidly instructed to go over it systematically mile by mile, missing no part. The major
difference from the year before was that, having lost
faith in Hedbäck's Point at the southeast corner of the
field, the search was started diametrically opposite at
the northwest corner, near where the British destroyer
had found the mail bag. This spot was also looked on
more hopefully because Captain Barzic (late of the S.S.
Seine which had sunk the *Egypt,* and who therefore
felt that he of all persons ought to know where he
had sunk her) swore she lay in that area and assisted
the *Artiglio* in her search there.

But in spite of all Captain Barzic's Gallic enthusiasm
in backing up his opinions, the *Artiglio* and her consort
found nothing in that northern area. Grimly they stuck
to their task, marking out with buoys a small section
of that vast checkerboard of ocean which they had to
search, sweeping it carefully with the drag, then moving the buoys to the next section to repeat the operation. Square by square they went over the ocean, two
small ships bobbing on the waves with a long line of
buoys dancing in a curve between their sterns to mark

out the line where, 370 feet below, that mile-long wire was groping for the *Egypt*.

The summer dragged along. The *Egypt* remained undiscovered. Captain Barzic was found to be a false prophet, as unaware of exactly where he had sunk the *Egypt* as had been the captain of that unfortunate ship. Endlessly, day after day, whenever the weather permitted, the *Artiglio* and her consort, 30 miles at sea, put out their sweep and worked over a new section of the sea, in the morning sweeping it from northeast to southwest, in the afternoon, reversing direction and going from southwest to northeast in the hope that, if the sweep slipped over the wreck in one direction, it would catch when going the opposite way.

In the midst of this, science was brought in once more. Another electro-magnetic finder, this time developed by Swedish electrical specialists of Stockholm, was brought aboard, together with the Swedish engineers; and again delicate electrical instruments were towed over the ocean, while Swedish engineers and Italian seamen gazed hopefully at flickering needles as the finders bobbed about the sea astern. But nothing resulted, and the Swedes and their electrical gadgets went back to Stockholm, while Quaglia and his men, disillusioned in everything but their own ability as seamen, went back to their sweeping.

June, July and August of 1930 slipped away. Two wrecks indeed they had found—one the Greek steamer *Demetrios Inglessis*, sunk during the war, and the other an unknown ancient wreck. Each, hooked hard by the

sweep, caused every heart on the *Artiglio* to beat madly
while a diver went down to see if it might be the *Egypt*.
After each proved to be a worthless hulk, with deadened spirits the sweeping was resumed.

August dragged away. The second year's work was
nearly over, and still no *Egypt*. The *Artiglio*, with
practically the whole area in which the wreck might
lie thoroughly covered that season, was nearing the
last corner of her rectangle for search, approaching
again Hedbäck's Point where originally, the year before, she had started work with the old sweep and
had caught so many rocks. Quaglia faced a dismal prospect—out of pocket the terrific cost of two years' search,
he was approaching the end of the possible area in
which he should have found the wreck, and there was
no wreck. Had he missed her, and would he have to
sweep the whole 60 square miles again? Or was the
Egypt perhaps after all completely outside the area he
had dragged? If so, on which side should he extend the
search, east or west, north or south? Or was the *Egypt*
perhaps lying on her side or even bottom up, so that
his new sweep slid harmlessly over her flat side or curving bottom, catching on nothing? It might be possible,
who knew? But still there were a few square miles to
be searched yet, and until those proved barren he had
no need of new decisions. The search went doggedly on.

Finally toward the end of August, in lat. 48° 6'
North, long. 5° 30' West, one mile west of Hedbäck's
Point, the sweepers caught something. Not too hope-
fully, in view of previous disappointments, a diver was

cased in an armored shell and dropped over the side to peer in the dim light of the deep sea at what they had caught. Pandemonium broke loose on the *Artiglio* when from 360 feet down the diver reported over the telephone:

"The *Egypt!*"

And so after two years' search and huge expense, they had found her. Deliriously on the decks of the *Artiglio*, they cheered and sang. The *Egypt!* Five million dollars in gold and silver lay there below them! They danced, they hugged each other, intoxicated by success after their long months of discouragement. Two years, but now they had her! Yet, had they known what lay ahead, there would have been no cheers. For compared to what was to come, what had passed was happy child's play.

CHAPTER XVII

FORTUNATELY they found the *Egypt* right side up, on an even keel, practically intact. She lay on a smooth bottom; unlike most wrecks, little damaged by currents during her long submersion; her stacks, her masts, her light superstructure still in place, all covered by a fine marine growth. For this good fortune, the great depth was mainly responsible. The ship was known to be on her beam ends as she disappeared from the surface, but as she flooded completely in sinking thereafter, the unusual depth gave her a chance to right herself again before she hit bottom, and, luckily for the salvagers, she had. The great depth had also saved her from the battering action of the currents, which had collapsed many another wreck.

But there the luck of the salvagers ended. For the bullion room was located three decks down in the hull of the *Egypt*. That strong room was inaccessible to divers in the Neufeldt & Kuhnke armored suit. In a flexible rig, an expert diver might easily have walked down the three flights of stairs leading from the superstructure practically to the top of the treasure room, or dropped down the narrow loading shafts, but in the armored suit it was out of the question. In that, a man could hardly shuffle along on the level; descending stairs or getting through doors or down the confined loading trunks leading to the strong room was impossible. There was nothing for it but to blast away the whole ship down to the plating over that bullion room so that the diver in his metal rig could be dropped vertically into the hole made as the blasting proceeded.

Carefully the plans of the *Egypt* were studied. The bullion room was a narrow athwartship compartment, 28 feet wide, 4.5 feet long, 8 feet high. A queer shape for a room—athwartships it was half the width of the ship; fore and aft it was very narrow, only a yard and a half; vertically it was a full deck height, nearly 8 feet from orlop deck to main deck.

To get into that room, the whole ship from side to side and for a fair distance fore and aft would have to be torn away down to the main deck; that is, a section of superstructure and steel hull 54 feet wide, 33 feet deep, and perhaps 60 feet long—an immense task.

Quaglia and his divers went at it. He had for divers three men, Gianni, Francheschi, and Bargellini, physical giants all. They had to be, to move about even slightly inside that armored suit when the deep sea started to squeeze the joints tightly together. Of these three, Gianni, chief diver, was also a mechanical genius, responsible for many of the queer devices used by Sorima in its salvage exploits.

But long before now, Gianni, Francheschi, and Bargellini had had enough of Neufeldt & Kuhnke's armored diving suit. Their long months of going up and down to explore wrecks (which had mostly turned out to be rocks) had convinced them of its trifling value, either in getting about on the bottom or, using the clumsy hooks protruding from its arms, in doing any work from it or with it. At most, it protected them only from the sea pressure while they peered out of its face-plates through the dim water. But if the rig were useful to them only as an observation case, they didn't need its joints, which were both nuisances and points of potential danger from leakage, or its claws, which might get entangled in wreckage. They might just as well abandon it altogether, and do their observing from a smooth cylindrical shell which had no joints at all and no projections.

And that they were ready to do. They had built themselves a cylindrical steel shell, somewhat resembling an Egyptian mummy case, with five glass ports near its top out of which they could see, thus getting back generally to the original conception on which the two English engineers (who originally started to design diving gear for the *Egypt*) had concentrated. Inside their shell, or "eye," they were wholly incapable themselves of locomotion, but at least they were comfortable and safe, and the ship above them could even more easily move them about and deposit them against a bulkhead on the wreck than it could the armored rig.

This observation shell reduced the diver to an observer only, as incapable of doing any work outside it as Beebe in his bathysphere; but over the telephone, at least, he could guide and direct the men from the ship above while they grappled, fished, used grabs and clamshell buckets, planted explosives, and in brief did the job from the surface, using the diver only as their "eye" on the wreck below.

In its previous sense, diving was thus completely abandoned. The salvage on the *Egypt* became the most gigantic fishing job on record.

And so they started. The divers said from their previous examinations that the wreck was the *Egypt*. But was it? Quaglia had to know. The divers might be mistaken. It was dark down there, they could not see far; more definite proof was necessary before the tedious work of blasting a wreck to pieces was begun. It would be a crushing blow to discover later that they were blasting on the wrong wreck. Something unmistakable had to be recovered to make sure.

For this purpose, nothing could be better (or more easily obtained) than the captain's safe. It would have unmistakable documents in it, and the captain's cabin was on the very top of the superstructure, directly exposed to action by grappling hooks from the *Artiglio*.

With a man encased in the cylindrical shell landed on the superstructure of the *Egypt* as an observer, the men on the *Artiglio's* deck went fishing below with a grappling hook. A queer sight. Against a bulkhead on the *Egypt's* superstructure leaned the motionless observation shell, its painted sides a ghostly white in the dim twilight of the depths. Inside it, nose and mouth covered by the breathing mask, pupils dilated in the semidarkness, staring out of the glass ports, was the diver. A few feet away, swaying gently in the current, bobbing up and down as the *Artiglio* rolled in the waves far above, was the grapnel, hanging from a cable at the end of the *Artiglio's* boom. The fishing commenced.

In a stream of staccato Italian, the orders of the diver were telephoned up.

"Forward a little! Down more! A fathom to starboard! Forward again!"

Obediently on deck the boom was trimmed in accordance with the orders relayed from below and repeated in hoarse shouts to the men at the winches. Slowly the hook, invisible in the water below to those who were moving it, groped its prongs toward the overhanging roof of the captain's cabin in the still depths, rubbed finally against the cabin side, scraping away the thin growth of seaweed there.

Excitedly from below came an order:

"Good! Heave in!"

The winchman threw in his clutch, reeled in the cable. The hook, bumping the cabin bulkhead, rose up and dug its prongs into the overhanging ledge. The *Artiglio*, heaving steadily, heeled down a little as the strain came on the cable, then the canvas-covered roof of the little cabin tore free of its light fastenings and came clear in one piece, exposing the room below.

The observation shell was hauled up a few fathoms, then dropped down again inside the unroofed cabin. There in plain sight in the water-filled room lay the captain's safe, a small steel cube perhaps two feet on a side.

Again from the surface the fishing commenced, this time with a large pair of jaws, for all the world like a huge lobster's claw. This was a harder fishing job. Directed as before, the men on the *Artiglio* groped round the cabin with their claw, trying to poise it over the safe, hampered mainly by the motion of their own ship which kept the jaws swaying unevenly on the cable. Patiently the observer below guided their efforts, each time the jaws came over the safe ordering them dropped sharply in an effort to seize it. Finally he succeeded. The jaws were snapped shut by a heave from above, gripping the safe hard, and, gently, the lift commenced. Squeezed between two steel claws, the little safe was hoisted to the surface, swung inboard, landed on deck. With that trophy in its possession, the *Artiglio* hauled up its observer, cast loose its moorings, and steamed for Brest.

There in the presence of numerous dignitaries from Lloyd's and the British Consulate, the safe was burned open with acetylene torches. Inside it was nothing but papers, but they were enough. Envelope after envelope proved to be diplomatic correspondence on its way from London to Bombay, consigned via the *Egypt* and for greater safety never entrusted to the mail room but given directly into the captain's custody. The British Consul promptly identified the confidential documents and took them in his charge. There could no longer be any doubt. The wreck at lat. 48° 6′ North, long. 5° 30′ West, was the *Egypt*. All that was necessary now was to blast a way through three decks down to the bullion room.

But the salvagers got little farther that year. Working now meant holding the *Artiglio* tightly over one spot on the *Egypt* while they fished below, with far less movement of the ship even than is permissible when working with ordinary divers. It was September, the weather at sea was getting bad, the *Artiglio* could not be held over the wreck. So the second season ended with the *Egypt* found and positively identified, and that was all. Nothing of value had been recovered. Sorima was out the cost of two years' work, and was still facing a staggering demolition job before it got a cent back. To help out his finances in the interim, Quaglia contracted with the French Government to demolish several wrecks which lay in the shallower and more sheltered water around Quiberon where they constituted obstructions to navigation.

The *Artiglio* and her consort, the *Rostro*, moved shoreward and started the work of demolition. The *Artiglio's* job was to cut down, to a point so low that any ship could pass over her at low water, the wreck of an American steamer, the *Florence H.*, sunk accidentally during the war off the island of Houat, near Saint-Nazaire.

The *Florence H.* was a ticklish wreck. Loaded with high explosives for the American army in France, she had lain submerged for thirteen years. Presumably long ago the waterlogged TNT filling her holds had become harmless, but of that there could be no certainty. So when, in September 1930, the *Artiglio* and her divers tackled the job of blasting away her upper works, they naturally enough tackled it gingerly. One after another, Gianni, Francheschi, and Bargellini went down on her in regulation diving rigs to plant the first charges, then moved the *Artiglio* a long way off before exploding them, lest the *Florence H.* go up in one vast detonation and blow them out of the water.

But except for their own moderate charge, nothing went off. Back to the wreck they went to plant a second charge. Again nothing happened. Gianni, chief diver, gained more confidence. The explosives on the *Florence H.* had evidently been too long in the water to be dangerous any longer. He ceased moving the *Artiglio* so far away between blasts.

So through October, November, and into December the divers worked, bit by bit tearing the *Florence H.* to pieces, good practice as they grimly noted, for their coming work on the *Egypt.* With professional interest, after each blast they scanned the damage done to the hull, noting how much steel each charge tore away, within what range the exploding powder acted. As the weeks went by and charge after charge was monotonously exploded, they forgot all about the *Florence H.'s* cargo, intent only on getting through before the weather got really cold.

December 7 came and the job was practically completed. Cut down almost flush with the mud line, the *Florence H.* was no longer recognizable as a ship. One more blast and the task was done. The last charge was planted, the *Artiglio* moved off as usual a short distance to fire it and as usual the firing circuit was closed.

Hell broke loose. Like a volcano erupting from the sea, the entire cargo of the *Florence H.*, thousands on thousands of tons of TNT, exploded with an earth-shaking roar. A vast cloud of smoke and spray rose skyward; the waters, as if thrust apart by a titantic hand, surged back, then in a huge wave rushed in to fill the void in the ocean caused by the detonation. When the smoke cleared and the sea had calmed a little, the *Artiglio* was gone. When the nearest vessel (fortunately for near-by ships, none was really near) reached the scene of the disaster, it picked up a few dazed survivors clinging to bits of wreckage.

Gianni, Francheschi, Bargellini, the three divers, were dead, together with the captain of the *Artiglio* and most of his crew. The seamen who for two heartbreaking years had struggled with the ocean to find the *Egypt,* and the divers who had gone to depths never before plumbed to get down to her and start the work of salvaging her millions, lay dead in the waters of the sea—not so far from those tons of golden bars and shining sovereigns and massive silver ingots filling the strong room of the *Egypt* which no longer meant anything to them.

CHAPTER XVIII

In the blast that destroyed the *Artiglio*, Quaglia lost not only his salvage ship but practically the whole organization he had built up since entering salvage work. Worst of all, he had lost the friends and associates of years.

There was nothing he could do to salve that wound, nothing he could do to bring the dead back to life, but one thing he could do for their memories, and that was to show the critics who had laughed at their efforts for two years that the job to which his lost assistants had given their whole souls could be done. With grim determination, he set about doing it.

Getting a replacement vessel for the *Artiglio* was not difficult. A similar vessel was soon purchased and, after a winter's hard work, refitted with salvage gear equal to the *Artiglio's*. But his divers he could not replace. Gianni, Bargellini, and Francheschi had been giants, men so large they filled completely the metal armor they had used for years. In all Italy there was not their like; it had required their brawny bulk to maneuver with any success the 700-pound armored shells in which they had originally worked. But there was still a chance. No longer were such giants required if he could make a success of the observation chamber which the occupant could not move anyway. So from the crews of his other vessels, mainly the *Rostro*, he

selected four substitutes—Raffaelli, Mancini, Lenci, and
Sodini—all smaller men and of far less experience in
diving than his lost trio. These, together with a new
crew, he assembled on the new *Artiglio*, and, at the
end of May, 1931, they moved out to continue the
work on the *Egypt*.

To hold his ship for working, he planted 6 five-ton
concrete blocks as anchors in a circle some 600 yards
in diameter centering over the wreck. To 6 buoys secured to these anchors the *Artiglio* always moored, with
6 wire hawsers, one to each buoy. So expert did the
·*Artiglio* become in handling lines that she could steam
into her circle of buoys and moor to all of them in
less than twenty minutes. When the weather became
so bad that she could no longer hang on against the
seas pounding her, she would let go 5 buoys, and, head
on to the waves, ride to the sixth. In thick weather,
to avoid being herself run down and sunk on top of
the *Egypt* (for she lay in the main traffic lane from
England to the Mediterranean and Africa) her radio
kept sending her position, and every three minutes her
wailing siren shrieked out a warning to near-by ships.
The nearest land was 30 miles away; the salvagers could
not, except by losing precious time, afford to run to it
for shelter, and consequently, through fair weather and
foul, the *Artiglio* and her company tossed in their little vessel on the bosom of the wide Atlantic weeks on
end.

So commenced the work of demolition. Working
with hooks, with grabs, and with claws, the *Artiglio*

fished from above while the divers in the shell below directed the fishing. Soon the light superstructure was torn away and tossed clear of the *Egypt's* side. Then came sterner stuff, the steel plates of the hull itself.

For this, blasting powder was used, in varied charges tamped into small metal cylinders, with the cylinders themselves lashed to a wooden framework of a size to cover the plates to be torn free. Here came in a problem requiring great skill and involving considerable hazard to the success of the venture—each charge had to be powerful enough to blast loose the adjacent steelwork, but never so strong that any explosion or any series of them should so jar the *Egypt* that the floor of the strong room, with 50 tons of bullion pressing it down, should give way and drop the treasure into the hold below, thus making its recovery vastly more difficult and probably even impossible.

The blasting commenced. Raffaelli, as befitted his position as new chief diver, was cased in the observation chamber and swiftly lowered into the depths till he was poised over the *Egypt's* hull. Then gently he was lowered the remainder of the way, delicately moved about from above as he telephoned instructions, till his shell could be dropped on deck leaning against a bulkhead to hold it erect. Never could he be left suspended in the water, for the motion of the *Artiglio*, communicated thus to the suspended diver, made observation by him then impossible. When finally Raffaelli had been landed to his satisfaction near the working point (within a few yards, because beyond that he could not see through the water in the dim light), his lifting cable was slacked somewhat, and the second step began. Down through the water toward him came the wood framework with its string of blasting charges. When he sighted it, the real job commenced. Carefully moved about from the ship above as he ordered, he watched it swaying up and down in the water till he saw it over the plating first to be removed. Then a quick:

"Good! Let go!" from Raffaelli, and on deck the winch was swiftly slacked off.

A last look to make sure the explosives had landed properly, and Raffaelli's job was, for the moment, done. Hurriedly, through 60 fathoms of water, he was hoisted up, his shell was opened, and he emerged to breathe and rest awhile on deck.

On the *Artiglio*, the electrician completed the firing circuit, and the switch was closed. A sharp shock vibrated the *Artiglio*. The charge had fired. What was the result on the *Egypt?* But in spite of a natural anxiety of the salvage crew to find out, there was no rush to close Raffaelli in the "eye" again and send him down to look. For the water below would be so roiled by the explosion that some time must elapse before it settled enough to make vision possible again.

Half an hour later Raffaelli went down to inspect. Landed again against a side bulkhead after some maneuvering to get him near it, he peered anxiously out of the port, his heart beating a little faster as he noted the results. The blast had been successful. Broken rivets lay all about; dished in amidships but curled up at the edges were several broken steel plates. He ordered the grab sent down, a heavy gadget with eight sharp steel claws which closed in on its center when its hoisting cable tautened.

Again the fishing started, the broken plates being the object. Time after time, under Raffaelli's orders, the swaying grab was moved over the edges of the shattered plates, dropped, heaved sharply up again. Finally the claws closed about a curled up edge, hung on, bent it up a little farther. Hastily Raffaelli was hauled aboard, and, between *Artiglio* and *Egypt*, a tug of war commenced.

The winch on the *Artiglio* groaned as its pistons spun the drum, heaving in the cable. More and more the *Artiglio* heeled over as the strain increased till her starboard rail was awash, and then the heaving stopped. She dared go no further lest she capsize.

Thus with her teeth in the *Egypt*, like a terrier with its jaws clenched on a bull, the *Artiglio* hung grimly on, a 15-ton strain holding the steel cable that disappeared into the water tense as a piano string. Below, rivets started to give way under the pull; as each one snapped, the *Artiglio* shivered a little, straightened a trifle. Finally, with a jerk, the remaining rivets tore loose and the *Artiglio* suddenly rolled wildly to port, oscillating back and forth until she came to rest, still heeled to starboard from the load hanging on her boom there.

Once more the winch heaved in till the grab came above the surface, clutching in its teeth a bent steel plate torn from the *Egypt's* deck. Hurriedly the *Artig-*

THE GRAB CAME ABOVE THE SURFACE, CLUTCHING IN ITS TEETH A BENT STEEL PLATE TORN FROM THE "EGYPT'S" DECK. (NOTE THE OBSERVATION SHELL, PAINTED WHITE, HANGING AT THE "ARTIGLIO'S" RAIL. FROM INSIDE THIS SHELL THE DIVERS ON THE BOTTOM DIRECTED THE FISHING OPERATIONS.)

lio was hauled to port, well clear of the wreck below, and the grab was opened, letting the deckplate fall with a splash back into the sea. The first opening had been made in the three decks shielding the bullion room. The *Artiglio* was moved back into position, another diver, Lenci this time, put in the shell, and once more the fishing recommenced.

So through June, July, August, and September, Raffaelli and his team-mates worked away on the hulk of the *Egypt*, groping with strained eyes in the murky water of the depths to place explosives and direct grabs, while, from the tossing *Artiglio*, her seamen, fingering the throttles of their winches, fished about in the depths for the wreckage, seized it with steel jaws, plate by plate tore the *Egypt* apart. The work went slowly, interrupted often by bad weather, but doggedly they clung to the task.

Fall came; the working days between storms grew scarcer, but they kept on. The boat deck, the promenade deck, the upper deck of the *Egypt* were all torn out, lying twisted heaps of junk on the ocean floor near by. Only the thick main deck lay now between them and the bullion room, and frantically they worked, as autumn slipped away, to get through that before winter put an end to their efforts. Five hundred tons of iron had been torn and blasted out of the *Egypt*, which lay now on the bottom with a yawning gap in her structure 100 feet long, 55 feet wide, and 33 feet deep. After each blast, tons of other debris—cabin furniture, bedding, innumerable fittings—fell from wrecked staterooms into the chasm and had to be removed before the broken steel was again exposed for grappling.

And now at last the main deck lay exposed, with the gold beneath its thick plates. Throughout October and November the divers worked between storms, praying for enough good weather to let them tear away those last few plates and recover a few pigs at least of all that gold before they had to quit. For three long years had gone by since Sorima started; well over half a million dollars had been spent on the job, and the sight of even a slight amount of bullion would have a tremendous effect on heartening everybody for the future.

But as they tore away at those last plates, they had to work gingerly. Now, if ever, the explosive charges must be small; just one violent blast and the bullion room might lose its floor. Should that disaster happen, Heaven alone knew where the money might come from, on top of the half million already gone, to finance blasting the ship apart down to its keel.

So with souls torn between a maddening urge for haste in removing the plates and a deadly fear of using too much TNT, the divers blasted and clawed at those last plates to get at the treasure. By the middle of November, they succeeded in tearing one plate off over the bullion room. The way to the gold was open at last. Cased in the armored shell, a diver was lowered down to explore that room, but it was too late in the season. With the sun even at midday always low in the sky, not enough light penetrated the depths for him to see anything or to permit further work. After one more

attempt on the next good day, December 1, which similarly ended in a failure to see, Quaglia gave up, and
work ended. Three years gone by and nothing recovered. Quaglia was personally out over $500,000. Would
he ever get it back?

CHAPTER XIX

MAY, 1932, arrived, and once more the second *Artiglio* was back on her station with faster hoisting winches and a rested crew. They had with them also a very special grab, a sort of orange peel affair that would go inside an opening a yard wide and, when its sections closed, would scoop together whatever lay beneath its jaws. But since what they hoped to recover was gold and silver bars and sovereigns, all small stuff, they took special pains that none be lost from the grab on the long lift through the water, to which end they provided a second clamshell grab 10 feet above the first. When the main grab and its contents were hoisted 10 feet, they nestled inside the clamshell, which thereupon closed beneath to form a tight receptacle into which anything which dropped from the orange peel would surely be held.

Immediately the task of tearing off the remaining plates over the strong room was commenced. Carefully, with small explosive charges, the final plates were loosened, seized in the grab, and torn away. The strong room was soon completely uncovered. With joy, Raffaelli, inspecting the space below, noted that the bulkheads had not collapsed; despite the five tons of high explosives he had used in blasting away the *Egypt's* hull, the room was still intact. Its bulkheads were bulged inward slightly, making a narrow room still narrower, but at least the room was still there and, to his surprise, nearly full to the top, whereas he had every reason to believe that the bullion in it took up not one third of its capacity. That was a puzzle to him, and even more so to Quaglia when he reported it. But the way to the strong room was open, and its contents were still there. Nothing else mattered. Hurriedly he ordered the treasure grab sent down to haul out the first load of gold.

Peering from his inboard port in the armored "eye," he soon saw floating into view before him in the water the clamshell bucket. Silently he watched in the unearthly stillness of the depths as the two halves of the clamshell opened wide, the inner grab dropped clear, and commenced to oscillate gently up and down, its four orange peel sections spread wide apart in the water.

Magically those hungry jaws began to move over the distorted main deck of the *Egypt*, searching for the narrow opening as Raffaelli, locked tightly in the observation shell, crisply gave his orders, and far above the winches on the *Artiglio* spun round and its boom swayed obediently to the will of the diver 60 fathoms down. The grab plumbed the thin athwartship black streak showing in the midst of the broken plates.

"Good! Let go!"

With a rush, down dropped the grab fairly into the hole, came to rest. Another sharp order and the four jaws snapped shut, crunching their way through what lay below. Then slowly the treasure grab rose up and disappeared 10 feet above into the larger clamshell. The two huge halves of that came together, and Raffaelli gave his final order:

"Hoist away!"

The grabs shot upward and in a few seconds disappeared from his sight, carrying to his shipmates above their first catch of the long-sought treasure.

Unsuppressed excitement gripped the *Artiglio*. As that grab broke surface and swung in over the rail, the entire crew gathered expectantly around the dripping buckets. Their heartbreaking search was at its end, the flow of gold was about to begin.

The outer clamshell opened, the grab dropped clear, hovered over the *Artiglio*. Another jerk on the winch throttle, the orange peel opened out and a rain of dripping yellow metal fell on the deck. Eagerly the cheering seamen rushed in, clutched at the golden shower.

And then their faces suddenly lost their smiles, their cheers choked in their throats. That yellow metal was not gold; those larger objects there were not ingots. Blankly they fingered what the drag had brought up. These were simply brass cartridges; that was only the waterlogged stock of a broken shotgun. None of it had the slightest value. Cold fear clutched at their hearts. What was such valueless cargo doing in a treasure room? Was it, after all, the treasure room they had reached by their years of toil and agony? Or merely some hold filled with worthless baggage like the trash before them? The jubilation suddenly died.

Listlessly Quaglia picked up the largest object, the shotgun stock, examined it. There he got his first sight of gold, but it gave him little comfort—it was only a gold nameplate set in the walnut, with the initials M.B.S. surmounting a princely crest—Mohinder Bahadur Singh, Maharajah of Patiala. He tossed it aside, waved silently to the winchmen to send the grab down and try again.

But the excitement, the anticipation, and the joy were gone for good from the faces of the men of the *Artiglio*. As the days grew into weeks, and, favored by exceptionally good weather, their grab made shot after shot into that room only to come up and pour out on their decks a varied assortment of worthless trash—innumerable cartridges, rifles, shotguns, miscellaneous baggage—their hopes died and their brows grew sullen. What kind of compartment were they rifling anyway—one of the many baggage and storage rooms? Had the divers in spite of all their care made a mistake and, at the cost of half a million dollars, blasted their way into a baggage room? It looked so. Over the waterfront in Brest, whence came their supplies, that rumor spread; thence it went to Italy, losing nothing in transmission. The *Artiglio's* men hesitated on their few trips ashore to show themselves—jeers and ill-concealed derisive gestures greeted them everywhere. Would they ever again be able to hold up their heads back in the little Italian seacoast villages whence they had come?

For a brief time the clouds lifted and gave them hope. Mixed with small arms ammunition, the grab began to bring up bundles of small watersoaked rectangular sheets, which on examination turned out to be brand new currency; 5, 10, and 100 rupee notes marked with the seal of the Indian State of Hyderabad. Soon they had over a million dollars worth aboard, littering their

decks everywhere, drying in the June sun. It was not
the hoped-for gold, but at least it was money. Quaglia
looked at it with interest. No mention whatever of those
banknotes had ever come to him in the list of treasure
the *Egypt* was carrying. Over a million dollars in rupee
notes! Perhaps after all they must be delving in the
treasure room, for where else would so much money
be carried? And his half of it would at least pay out
his costs, even if they never found the gold.

But swiftly disillusionment followed. In reply to excited radiograms sent to London reporting the recovery
of the rupees, came the chilling answer that the notes
were worthless—that immediately upon the sinking of
the *Egypt* the State of Hyderabad had canceled them
all and had printed and put into circulation a replacement issue. And besides, they were informed that an
essential signature, that of an Indian dignitary, was missing from those rupee notes, printed in England, which
had never reached India for that signature. Both London and Bombay agreed that the rupees which Quaglia
and his men had spent precious diving days in fishing
up with their grab were simply beautifully engraved
waste paper, which might just as well be burned as fuel
to save a little coal.

Once more gloom fell on the *Artiglio*. So that paper
money meant nothing in itself; its presence in the room
they were fishing in meant nothing either. Unsigned
notes, not specially insured for their face value, might
be going out simply as express matter, stowed wherever
in the baggage rooms there was space to jam them.

But Quaglia did not agree as to the worthlessness of

100 RUPEE NOTE, OF THE INDIAN STATE OF HYDERABAD, FISHED UP FROM THE "EGYPT." NOW
RUBBER STAMPED AS "OF NO MONETARY VALUE"

those banknotes. Carefully he had all the rupees gathered up and locked safely below. Money was money,
wherever found, and if all else failed neither Britain
nor Hyderabad was going to repudiate it so far as he
was concerned without a legal struggle. Nor did he lose
hope in spite of all the discouragement about him on
the *Artiglio* and the sarcastic jibes flying about ashore.
He had firm faith in his divers. They had followed his
directions in ripping apart the *Egypt*, and the room
they had reached *was* the bullion room. Why that room
was so jammed to its top with miscellaneous junk which
could by no means rationally belong in a treasure room
he could neither understand nor explain, nor did he try
—it was simply one of those idiosyncrasies of the English
who were incomprehensible. But the bullion was aboard;
for that he had Lloyd's word and indisputable evidence
in bills of lading. Under all the rubbish 60 fathoms
down *must* be that treasure. He kept his dispirited men
fishing in the room they had uncovered. There would
be no blasting in any new compartment until that room
had first been emptied down to its bare deck.

The monotonous labor of fishing with the grab went
on. A month had gone by since the *Artiglio*, her crew
full of hope and eagerly expecting soon to see treasure
showering down on deck from the new grab, had
started her fourth year's work. Now that hope was
gone. Apathetically the men on deck watched as each
shot of the grab deposited more rubbish on their decks.
Then on June 22, came a change. A seaman, perfunc-
torily rummaging through the latest mass of soggy
debris dumped from the jaws of the grab, came upon

two shining golden sovereigns! He held them aloft excitedly. They were worth only $10, but they were gold! It began to look as if perhaps after all they might be working in the treasure room.

Like magic the entire crew of the *Artiglio*, except a solitary fireman tending the boiler below, gathered round the boom as the grab went overboard and the wire cables sang on the sheaves while it sank swiftly down to hover again over the *Egypt* far below. Raffaelli, chief diver, poised at the rail, signaled the winchman as he moved the grab slowly about in answer to his diver's orders. In an agony of suspense they waited while the swaying grab in the depths was gently maneuvered about by a dilated pair of eyes, staring at it from inside that armored mummy case. Finally it hovered again over the gap in the *Egypt's* main deck.

"Good! Let go!"

Down dropped the grab, closed its jaws, and, to the rattling of the winches, started up again. Hardly daring to breathe, the men on deck watched as it broke surface and swung inboard. The outer clamshell opened, spraying water on everyone, while the heedless Raffaelli, tensely silent, signaled by opening his clenched fists to drop the load. Now they would see.

The orange peel jaws spread apart, dropped the usual load of trash. But that was not all. With a clatter, two heavy bricks rattled down on deck to lie there gleaming in yellow splendor amongst the debris!

Immediately a wild cheer rang out over the sea, releasing the pent-up emotions of the strained seamen.

"The ingots! The ingots! Gold! Gold!" and with an

unbridled rush the *Artiglio's* seamen flung themselves
upon those two golden bars, caressing them lovingly,
passing them from hand to hand to be eagerly felt, ir-
refutable proof at last that the treasure was theirs! Pandemonium took charge as the long-suppressed feelings
of the salvagers lost all restraint—tears, kisses, hugs were
bestowed indiscriminately on each other and on those
precious bars of gold. For a few minutes, the *Artiglio*
was a madhouse of gesticulating, shouting sailors.

Then Quaglia took charge, held his arms up for silence. Were all hands there? No. Below was one fireman; in the depths still the diver. Swiftly both men
were brought up on deck to join their comrades. Then,
led by Quaglia, all bowed their heads in memory of
their dead—Gianni, Francheschi, Bargellini—those giants
who first had led the way to that treasure of the *Egypt,*
and to their other shipmates of the first *Artiglio,* all of
whom they felt must somewhere be looking down on
them, rejoicing with them in their success. A moment
they stood thus reverently uncovered; then looking upward toward the masthead where suddenly a huge
Italian flag had fluttered to the ocean breezes, they
roared in unison:

"Viva l'Italia!"

And then they went to work again. Down into the
sea went the diver, after him the grab. Each shot at the
treasure room now brought up ingots and showers of
gold coin. When, on June 25, three days later, the
weather forced a suspension of work, they had on board
bullion to a total value of nearly $1,000,000!

CHAPTER XX

Lloyd's insisted that the gold be landed in England. So to Plymouth, from which Sir Francis Drake had sailed to ravage the treasure ships of Spain in the great days of Elizabeth, the *Artiglio* sailed with the treasure of the *Egypt*. Awaiting Quaglia on the dock, ready to congratulate him on his skill, his courage, and his success, were all the higher officials of Lloyd's, the local dignitaries of Plymouth, and the British Admiral commanding the Channel Station. As the flag-bedecked *Artiglio* steamed in, the dock was alive with cheering admirers, waving heartily to Quaglia on the bridge. But when the ship tied up and Quaglia came down on deck, the first person over the side to greet him was none of his admiring friends but a sheriff with an attachment to seize not only the *Artiglio* herself but all the gold she carried!

That Quaglia was stunned by this unexpected welcome puts it mildly. What was it all about?

He found out soon enough. One of the French expeditions that six years before, on some previous understanding with Lloyd's, had searched unsuccessfully for the *Egypt*, now that Quaglia had found her and recovered part of her treasure, had secretly sued out an attachment against him, claiming a share in the bullion. And there, with the papers loaded down with formi-

dable seals to enforce it, was the sheriff seizing both his
ship and the gold! What a greeting!

In shocked silence Quaglia turned from the sheriff
to Sir Percy Mackinnon, president of Lloyd's, who was
trying to edge past that officer of the law. Sir Percy
tried to soothe him. Lloyd's would put up the necessary
bonds, see that ship and treasure were released, and fight
the case in court. And so that much was done, but the
edge was completely off the *Artiglio's* celebration. With
much trepidation, Quaglia sailed again to resume work.
Would the harpies of the law finally rob him of the
gold that he had snatched after four years' labor from
the depths of the sea? Who knew? After all, in England he was a stranger in a strange land, and months
would pass before he would know what England's laws
would do to him.

The *Artiglio* went to Brest for coal and supplies before returning to her diving station. But between bad
weather and a sudden call to help out the French Navy
on an emergency diving examination of the French submarine *Prometheus*, sunk near Cherbourg, it was not
till over a month later, on August 8, that the *Artiglio*
could again moor over the *Egypt* and send down her
observation shell and her grab.

In all, five times only that year, between late June,
when the first gold came up, and early November, was
the *Artiglio* able to moor over the wreck, for a total
working time of 188 hours. Her longest stay over the
wreck was eight days in late August. The weather that
season was continuously bad; on November 3, 1932,

Quaglia had to cease for the year. But when for the last time, the *Artiglio* sailed that fall for Plymouth, she had recovered 865 bars of gold out of a total of 1,089 on the *Egypt;* she had retrieved 83,300 gold sovereigns, and hoisted aboard 6 tons of silver—to a total value of $3,700,000. And on his last trip into Plymouth, Quaglia learned to his intense relief that the claims of the French, after a bitter three months' legal battle, had been dismissed by the Admiralty Court as without merit, and the impounded gold was now released to him.

In 1933, the fifth year, the *Artiglio* went back to recover the remainder of the treasure, consisting now mainly of some 37 tons of silver bullion at the bottom of the strong room. She got most of it, but another year, 1934, was spent in a final cleanup, and not till then, six years after the start, was that $5,000,000 in gold and silver, sunk in the *Egypt,* finally retrieved by the second *Artiglio*.

And so ended the salvage on the *Egypt,* the deepest recovery job ever undertaken by men under the sea, a success due to the indomitable courage and skill of Commendatore Giovanni Quaglia of Sorima, backed up by the marvelous seamanship, the technical ingenuity, and the fortitude of the Italians under his direction.

Six years' work, twelve lives, and nearly a million dollars in cost went into that task. It was no job for amateurs. Only the firm faith that if they could crack apart the *Egypt* before they cracked themselves they would assuredly find the treasure there kept Quaglia

and his men at it till fortune finally showered them with
a veritable rain of dazzling gold and silver ingots.

There was one more tragedy, for me at least, connected with the *Egypt's* gold. Lieutenant Cuniberti,
Italian Royal Navy, had first informed me of the expedition, and as it progressed had occasionally kept me
in touch with what was going on, though, after the
first year, his duties in Italy had called him back there
where he continued his deep diving experiments. In
1932, while from 360 feet down the ingots were beginning to come up from the *Egypt*, he was startling
his late shipmates on the *Artiglio* by going down himself off Portofino to a depth of 650 feet in a special
diving shell.

In 1936, I was in Rome. Desiring to thank Cuniberti
for his previous information and perhaps to learn more
about his later experiments at great depths, I wandered
over to the Italian Ministry of Marine and inquired for
him.

I made little progress at first, for my Italian was exceptionally poor, but soon an officer, a commander,
was brought in who spoke excellent English. After in-
troducing myself by means of a letter which I had procured from our own Naval Attaché in Rome, I got
down to business. Where might I find my friend Lieutenant Alberto Cuniberti, Italian diving expert?

"Cuniberti?" The naval officer before me looked at
me in astonishment. "Why, Lieutenant Cuniberti is
dead!"

"Dead!" I repeated, shocked. "Cuniberti dead! Since when, and how?"

"He died a few months ago, in a diving accident. We were all much saddened; he was a fine officer. You had not heard then? He was a friend of yours? We have in our records a report on his death. If you wish, I shall send you a copy of the report through your embassy here in Rome. Would you like it?"

Sadly I nodded and left. So I should never meet Cuniberti and thank him for all his courtesies. But what had happened to him?

Next day our Naval Attaché sent me a report transmitted through him from the Italian Ministry of Marine. As I read, I could hardly believe my eyes. Cuniberti, who had dived several years before to 650 feet, had been killed in less than 6 feet of water! It seemed incredible, but there were the facts.

A few months before, off Portofino, he had taken out for trial a newer armored rig with the intention of testing it to a depth even greater than the 650 feet he had already reached. But bad weather prevented his going overboard, so as stated in the report:

"He postponed the experiment for ten days and meanwhile went to Turin to continue his practice and studies at the local swimming pool and at the same time train some pupils to dive with submersion masks. This apparatus had been tested by Professor Herlitzka and other persons, and consists of a kind of airproof hood covering the head and neck.

"To the hood is attached a small reservoir containing oxygen at high pressure. The oxygen is released into the hood by turning a key, thus insuring respiration. This artificial feeding of breathable air can continue for over one and a half hours. This mask has been found very useful in leaving submerged submarines and has never given any trouble whatever.

"Cuniberti, who had some pupils with him, dived in, wearing the mask, at about 2:40 P.M. As usual, he swam about on the surface for a minute and then let himself sink to the bottom, lying down on the cement floor of the tank in a place where the water is less than six feet deep. This was an experiment he often made, often remaining in this position for a quarter of an hour, and therefore the pupils felt no anxiety at seeing him remain motionless for some minutes.

"It was only after some time that one of the swimmers approached Cuniberti and noticed that he gave no sign of life. Without delay he freed the officer of the weights which held him down and brought him to the surface. On taking off the mask it was seen that his face was already of a deathly color.

"Doctor Aglesio, on duty at the pool, immediately gave an injection of adrenalin and performed artificial respiration for a long time. But everything proved in vain. Cuniberti had died of suffocation."

What had happened? Nobody really knows; something queer had occurred to knock Cuniberti suddenly unconscious, and he had suffocated in the sight of numerous onlookers without the slightest indication of his deadly peril, and before he ever turned on the oxygen to feed his breathing. Probably he had used up the little oxygen inside his mask and lost consciousness before opening up his oxygen cylinder.

So, ironically, in 6 feet of water, perished the man to whom deep diving was a passion and who first brought to my then incredulous eyes the information that he was going out with his countrymen to salvage the *Egypt's* gold, over 60 fathoms down.

CHAPTER XXI

THE year 1917 had just dawned. For thirty months the war had raged in Europe, and German submarines had taken heavy toll of British commerce. And now, to cap all, Germany had just announced an unrestricted submarine campaign with every ship, neutral or belligerent, approaching the British Isles to be sunk without warning. In the offices of the Admiralty at Whitehall strained faces looked at the mounting curves of tonnage sunk by U-boats and knew that the threat was no idle boast—Germany had enough submarines in operation to bring England to her knees in a few more months if sinkings continued at the current rate. Regardless of what happened on the Western Front or on the Eastern Front, the war would be lost at sea unless a curb on U-boats' successes was soon found.

For hard-pressed Britain, with her man-power drawn away from farm and factory to hold back the German hordes pressing across France, was vitally dependent now upon the steady flow across the seas to her of those supplies she could neither raise nor manufacture in sufficient quantity—American wheat and cotton, American steel and powder. But the getting of them to Britain was as much a headache in the Exchequer as it was in the Admiralty, for somehow those purchases in neutral America had to be financed or they would never become reality enough for Britain's sea lords to have to worry themselves over in safe transportation through submarine-infested waters.

Every device known to finance in London had already been used—British balances in the United States had been exhausted; the maximum that Britain could borrow in America had been raised; American securities owned by British investors had been mobilized in London and shipped to New York to be sold to provide further credits—but still all this was not enough. If the pound sterling was not to drop against the dollar in New York to the point where it would sadly cripple further purchases, gold must be shipped in huge quantities to bolster up the pound.

For some time for this purpose, the tide of gold had been flowing westward. Now, in January, 1917, another huge shipment was to start, and the Lords of the Admiralty were confronted with a major problem. How should they route that priceless gold shipment to escape having it sunk at sea by hidden U-boats?

To start with, they chose R.M.S. *Laurentic*, a large White Star liner, as the carrier, a vessel fast enough to outrun any submarine which sought to chase her, a vessel already converted into an auxiliary naval cruiser and heavily enough armed with naval guns to fight off any submarine which might attack her, a vessel already manned by a naval crew as skilled as any in detecting lurking periscopes among the waves.

Next they sought the safest route westward. The seas to the north of Ireland were not so much traveled and were usually too boisterous in winter time for U-boat

commanders to lie in wait for victims—the chart of
sinkings, emphasized by the point of loss of the *Lusitania*, indicated that the waters south of Ireland and in
the Channel were the favorite hunting grounds for
lurking submarines. So for the course of the *Laurentic*
they chose the route around North Ireland.

From London to Liverpool ran a special train with
3,211 ingots of gold carefully boxed up for shipment
on the naval auxiliary cruiser *Laurentic;* 3,211 ingots of
gold then valued at £5,000,000 or $25,000,000—in
present dollar value, about $44,000,000. In boxes weigh-
ing 140 pounds each, the gold was carried aboard the
Laurentic through one of the entry ports low down in
her heavy steel hull, carted athwartship to the second
class baggage room, and there locked up, 43 tons of
gold. In secrecy such as surrounded the movements of
every naval vessel, the *Laurentic* sailed from Liverpool
with her precious freight and little else aboard other
than her crew.

A short run to the northward through the protected
Irish Sea, and the *Laurentic* pointed her bow westward
through a wintry January gale, plunging heavily into
head seas as she fought her way into the Atlantic with
the near-by north coast of Ireland looming up on her
port hand across the tumbling seas. The *Laurentic's*
captain had good cause to congratulate himself on his
luck; in that weather no submarine, even if present,
could hope to make a successful torpedo attack, and he
would shortly be out on the deep Atlantic where submarines rarely went.

Then off Lough Swilly came disaster. A terrific explosion rocked the ship. No submarines were near, but one had been in better weather beforehand, planting mines beneath the surface. The unfortunate *Laurentic* struck a hidden mine which inflicted damage far worse than any torpedo might have done, and, heeling over, started to sink rapidly. As best they might in the freezing weather, the crew struggled to launch the boats from their foundering vessel. Between the icy water, the driving seas, and the heavy list of the stricken liner, they were none too successful in getting away, and when the *Laurentic* disappeared, over half her crew, several hundreds of seamen, went with her.

When from the frozen survivors who finally made shore in the ice-coated lifeboats the news of the loss of the *Laurentic* was radioed in code to London, deep despair struck the Exchequer and cold fury reigned in Whitehall. The Lords of the Treasury looked in anguish toward the Admiralty— £5,000,000 in gold gone, the largest sum in all history to be lost at sea. But it was worse than the mere loss of £5,000,000. The loss was in gold, and gold above all else was needed in New York to bolster up the tottering pound sterling and keep the stream of food and steel flowing toward Britain. Could the Admiralty do anything?

If it was humanly possible, the Admiralty would and could. The gold had been lost while in the hands of the Navy; it was the Navy's obligation to recover it.

To Commander G. C. C. Damant, R.N., the British Navy's premier diving authority, was given the task, with orders to start immediately. He was not to regard his task as a salvage job but as a military undertaking of the utmost urgency. It was wintertime, it was wartime, but, in spite of the weather, in spite of the war, Damant was turned to at once with the best divers the British Navy could muster under his command.

Minesweepers soon located the wreck. They found the hulk not far off the mouth of Lough Swilly in a depth of 132 feet of water, or 22 fathoms, and buoyed the spot for Damant. It was a terrible place to work. Unsheltered by any land, the *Laurentic* lay in the open sea, exposed to the full sweep of every northerly or westerly Atlantic gale, and with Lough Swilly to the southward to allow a stiff sea built up by any southerly gale to strike full force over the wreck. Racing tides sweeping back and forth along the coast meant fierce currents to be encountered by the divers. To top off all, January in that latitude meant freezing spray on deck to hamper the workers and icy cold in the depths below to numb the divers.

Then there were the Germans. It was unduly optimistic to hope that the loss of the *Laurentic* and her fabulous cargo could long be kept secret from Prussia's prying intelligence agents. What the prize they had struck from Britain's grasp meant in the way of munitions would be as well realized in Berlin as in London; the military value of frustrating recovery of the gold would be immense; and Damant could expect, if Germany found out what he was about, to have submarines sent north for the specific purpose of blowing his little diving ship, a perfect anchored target for a torpedo, completely out of the water, with himself and his men sure to be killed by the explosion without chance of escape.

With all that in the back of his mind, Damant went at the job. His first diver down found the *Laurentic* lying in the sand on her port bilge, heeled over about 60°, so that walking on the deck was wholly impossible and even clinging to the sloping starboard side was out of question except where the diver could grip a porthole, a padeye, or some other protruding fitting. But nevertheless with lead-soled boots braced against rivet heads on the shell, and freezing fingers clutching at what they could grasp, the divers crawled over the badly listed side, looking for the entry port through which the gold had been loaded. Amidships in the second class baggage room, down what would now be a steeply inclined passageway leading from that sealed entry port, they would find the gold.

Long ocean swells left by the last storm were sweeping in steady succession over the *Laurentic's* tomb, each wave, even in the depths below, setting up a strong surge as its crest passed. The divers, searching in the dim depths for the entry port, found themselves nearly swept from their precarious perches on the *Laurentic's* side by each surge, and were forced to cling tightly most of the time to the nearest fitting, scrambling along a few feet only between pulsations. To make matters worse, the *Laurentic's* crew in abandoning ship had of course lowered all the boats they could, and the boat-falls now hung down the ship's side to what had once been her waterline, with the heavy blocks at the lower

From "Deep Diving"; Courtesy Sir Robert H. Davis

THE "LAURENTIC" AS SHE WAS

ends of the falls swishing erratically with every surge like huge pendulums from the davits overhead.

To see, as the first diver clinging by his fingernails to the side did, one of those massive boat blocks go flying through the water within inches of his faceplate, nearly braining him, was a sight to chill any man's blood and give an impressive demonstration (if any had been needed) of the wave action going on below. Damant had first of all to cut loose those death-dealing boat blocks before his men could proceed in such reasonable safety as even divers are entitled to.

Another diver soon found the entry port and tied a buoyline to it. The mooring buoys of the diving ship were then immediately relaid about this marker as a center, so that the diving ship might plumb the hatch below and minimize the danger of fouling a diver's lines.

Damant (who was himself as fine a diver as any of his men, and as a lieutenant had taken a leading part in developing the theory of stage decompression), having checked all to his satisfaction, now went further. A charge of guncotton was placed against the heavy steel entry port doors and exploded electrically from above. His next diver found the doors torn loose as expected. As was not expected, the doors were found resting a few feet inside the ship against something, still a total obstruction to entrance. Only with some difficulty were the doors torn out of the ship and removed, to expose behind them a heavy latticed iron gate across the passage, against which they had been resting. Another charge of guncotton took this gate off its hinges, but it required two more dives to remove some heavy packing cases in order to clear the passage inboard to the strong room.

So far, in ordinary weather, what had been done by the divers might easily have been done in a couple of days, but Damant had been struggling to cling to his moorings in a continuous series of midwinter gales punctuated by snow squalls, and two weeks had gone by during which diving was possible only for brief intervals and even then under conditions when normally it would not have been attempted. But each time the wind and sea lulled enough to make it seem probable that the moorings would hold his ship even for an hour, down went a diver into the freezing water, and so, bit by bit, over a fortnight, Damant had managed to get done two whole days' work and he began to breathe a little more freely.

He had not yet been torpedoed, his men were doing well in that cold water, and the backbone of his job was broken—the way to the treasure room was cleared. It was a difficult way, down a passage sloping at an angle of 60° and around several corners, but Damant had often seen worse. It looked now as if a few weeks' work would see the 40 odd tons of gold in the *Laurentic's* strong room lifted box by box to the surface and shipped safely back to the vaults of the Bank of England—two weeks' time if he were favored by the weather, somewhat longer if he were not, but at any rate long before an inkling of what he was about got back to Berlin to bring a U-boat to the spot.

In the late afternoon of his fourteenth day over the

wreck, the last obstruction in the passage was removed. To Diver E. C. Miller, one of his best men, Damant gave the task of getting into the strong room. With a chisel and a short-handled sledge hammer lashed to his belt, Miller in the gathering darkness of a brief February day went over the side of the salvage ship, slid 63 feet down the descending line to the opened entry port on the high side of the listed *Laurentic*, and crawled into the ship. With his tenders carefully paying his lines out foot by foot, Miller half slipped, half dropped down the now nearly vertical deck going inboard, groped his way in the utter blackness of the water-filled passage around several bulkheads, and then felt out the steel door leading to the strong room. With sledge and chisel he smashed his way through the nearly horizontal door, to slide immediately into the black water inside the strong room and bring up sharply with his lead boots clattering on a pile of bullion boxes jumbled in a huge heap against the port side bulkhead of what had been the second cabin baggage room.

Never before or since in history has it been given to any diver to land on such a hoard of gold. Forty-three tons of it in 3,211 bars lay in a scrambled heap against the bulkhead where it had been tossed when the crazily heeled *Laurentic* hit bottom. Everywhere Miller reached out a canvas-clad arm through the black water he felt stout boxes of gold—$40,000 worth of it in each box, $25,000,000 of it altogether. Excitedly Miller telephoned to Damant that he was in the bullion room and that, except for the smashed door, the bullion room was intact and the treasure all there! He was more entitled than he then knew to his excitement, for Miller unfortunately was the only man who ever saw (or rather felt) that bullion all together in the sunken *Laurentic*.

Miller's task, forcing open the strong room, had taken all his diving time. He had been down an hour already; it was dark on deck; the sea was none too good for holding on. Sharply they signaled him to come up so the ship could unmoor. But with all that gold about him, Miller was not going to come up empty-handed. He seized the nearest box of gold, a small box about a foot square and six inches deep, and got, as everyone gets when first he grabs a golden ingot, an incredible shock at its weight. The little box weighed 140 pounds, no easy load even for a strong man to carry under the best of circumstances, and Miller's circumstances for carrying anything were atrocious. But he refused to let go the gold; he had first found it, he would be first to bring some up. So vertically up through the strong room door, around the bulkhead corners he struggled in the water-filled passages with his golden ballast, little help possible from his tenders above on the salvage ship because of the many turns and twists his lifelines took inside the wreck on their way down to him. Pushing the five ingots in that box ahead of him, he wormed his way up the steeply sloping decks with superhuman strength and agility, till finally in the clear his tenders got a straight pull and heaved him up to the entry port where at last he was able to lash his precious burden and send it up on a line. He himself spent the next half hour dangling at various stages in the icy water while he was hauled up in a much-shortened decompression.

So after only two weeks' work in the *Laurentic*, Miller sent the first gold up, but the strain and the excitement must have been too much for his circulation; he had soaked up so much nitrogen that within an hour he was being jammed into the recompression chamber for treatment for "the bends." Bubbles of air had gathered in his joints, doubling him into knots with pain. It was remarkable to note, however, when, inside the tank, the pressure had been run up on Miller to 20 pounds, how suddenly his pain disappeared, after which he was gradually decompressed for several hours down to atmosphere and then emerged from the tank feeling quite all right in his joints and much elated over his success.

When morning dawned again on the gray sea tumbling over the *Laurentic*, the barometer was falling. Commander Damant, eying it uneasily, concluded that with luck he might get in one more dive before he had to let go his moorings and run before the rising storm. That meant that if any more gold was to be recovered, only Miller, who already knew the way, had a chance of getting into the strong room quickly enough to have time remaining for any useful work; so overboard, in spite of his bout of the night before with "the bends," went Miller again.

Miller quickly demonstrated that he had learned the tortuous way into the bowels of the badly listed ship and profited by his experience of the previous evening, for in one dive of only 60 minutes, he managed to mule three more boxes of gold up out of the strong room and send them to the surface, a feat which to some degree took the sting off the imperative necessity of letting go the moorings immediately upon his coming up. With four boxes of gold, about $160,000 worth, in his hold, Damant ran for shelter into Lough Swilly before a mounting northerly gale, with every expectation of coming back to finish lifting out the rest of that $25,000,000 within a few more weeks. Had anyone then told him that he was to be at it yet for seven more years, Damant would have been completely incredulous. But so it was.

For a solid week a fierce winter gale blew from the north with ever-heightening storm waves sweeping in never-ending succession over the grave of the *Laurentic*, 22 fathoms down. Long before that storm blew itself out, the north coast of Ireland for miles around was strewn with wreckage from the sunken *Laurentic*, much of it easily identifiable as having come from the inside of the ship, an ominous portent to Damant and his men of what those waves were doing on the bottom to their wreck.

In deep trepidation when the storm finally moderated and diving could be resumed, Damant watched as the first diver went overboard to secure a new buoy line to the entry port. And his heart sank as he noted that the pressure on the diver at that entry port now showed it at a depth of 103 feet, whereas before it had been but 62 feet from the surface. Somehow the side of the

Laurentic was now 40 feet lower than it had been before the storm; he soon enough found out why. The
diver going through the entry port into the passage
below could get but a few feet; the deckplates forming the ceiling of that passage were squeezed down to
within 18 inches of its floor, and buckled bulkhead
plates jammed in between completely sealed off what
little space was left. Under the endless pounding of the
storm waves beating against her tilted sides and decks
as she lay far over on her port bilge, the *Laurentic* had
folded up like an accordion, and in way of the entry
port, even an eel could no longer squeeze through that
flattened-out passage to the strong room!

To Damant and his divers, that was a body blow,
but they had to get through again to that strong room
in the wreckage of the *Laurentic*, and they set out to
do it. With successive small charges of guncotton exploded in the crumpled passage, they forced apart the
steel plates, shoring up as they went inboard to make
a tunnel through which a diver might crawl. It was a
terrible job, with the broken plating overhead, five
decks of it, groaning and creaking and working like a
thing alive as the surging waves beat down through the
depths; and the diver, alone and in darkness, stretched
out in that quivering mass of steel supported by nothing in particular, well knew, as he wormed his way
along with guncotton or with shores, that if those plates
should fold up on him no one could ever get him out.

That it could ever have been done had not Commander Damant been himself a diver, I very much
doubt. But where the leader will go, the men will fol-

low, and the tunnel advanced till at last the way was cleared to the strong room and once more on deck they prepared to hoist out the gold.

For the third time, Miller slid down into the strong room, reaching it now at a depth of 120 feet, right down on the sea floor where the collapse of the *Laurentic* had dropped it. But this time as he slid into the inclined room, Miller brought up against the far bulkhead with a metallic clatter as his lead-soled boots landed directly on the steel and no heap of stout wood bullion boxes broke his fall. Anxiously he felt about him but it was useless—the bullion room was completely empty!

In a daze, Miller groped through the water, his numbed fingers traveling over deck, bulkheads, and ceiling, searching for that pile of gold, but he found none. Instead, gaping rents in the steel deckplates and in the lower bulkhead showed only where it had gone. In the widespread collapse and flattening out of the *Laurentic*, the strong room plating had given way under the load of that 43 tons of gold and had torn wide open, spilling the precious bars downward somewhere into the general tangle of wreckage of the ship to port.

For Damant and his salvagers that was a heart-breaking discovery. Gone now was every hope of quick recovery of that treasure which a few weeks before had indeed been within their grasp—the relentless sea seemed to have taunted them with the feel of it only to snatch it then abruptly far beyond their reach and bury it beneath thousands and thousands of tons of broken steel down in the sands of the ocean floor.

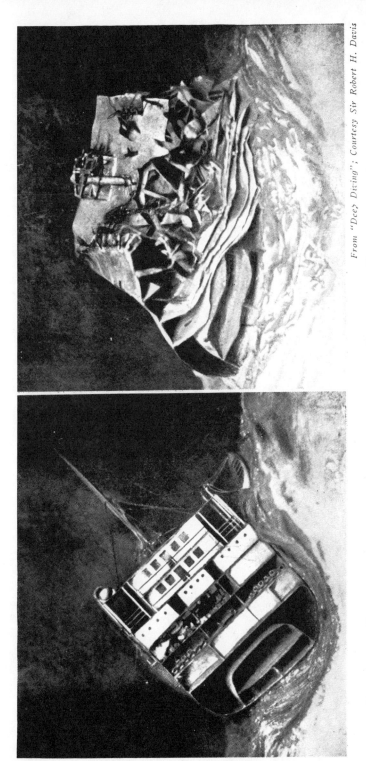

From "Deep Diving"; Courtesy Sir Robert H. Davis

THE "LAURENTIC" AS THE DIVERS FINALLY LEFT HER

THE "LAURENTIC" AS THE DIVERS FIRST FOUND HER

On the surface, Damant gloomily recast his plans. He had to recover that gold as his bit in backing up his shipmates of the Grand Fleet and his khaki-clad countrymen struggling to hold their own in the trenches on the Western Front. His enemy was as powerful and as ruthless as any they faced; but his weapons were slight in comparison, simply the weak bodies of a few men struggling with him under the sea, held to their task by no hope that any share in the *Laurentic's* gold would ever be theirs for their efforts.

It was obvious that the tunnel driven from the entry port in the starboard side of the ship down to the strong room was too dangerous for use in the gigantic task that lay ahead, and that, in spite of the peril and the labor spent in clearing that tunnel, it must now be abandoned. Nothing remained except on the bottom of the sea to tear the *Laurentic* to pieces, plate by plate and beam by beam, working vertically downward from her upper deck into her hold, till they came on the spot in the port bilge into which the gold must have been spilled when the strong room gave way.

With explosive charges, Damant began to blast his way down through the collapsed wreckage on the port side. To his despair, he quickly discovered that the loosely lying steel plates failed to break under the action of his explosives in the normal manner. With no strain any longer on those steel plates, they simply flopped loosely up or down under the impact of the exploding guncotton like flags waving in the wind, without ever parting. To sever the plates he found he had to seize each plate with clamps, heave taut on the

clamps with a line to the boom on his salvage ship until
the plate was strained hard out, and then fire a charge
under it to cut it free at its lower edge. Thus the work
went slowly ahead as, one after another, the *Laurentic's*
plates were blasted out and dumped well clear of the
ship. But not wholly without mishap.

Blachford, veteran diver, was working below.
Twenty fathoms down, clutching gingerly a charge
of guncotton with its detonator imbedded inside,
Blachford crawled on hands and knees over torn steel
through the water to get beneath a steel plate swaying, at its outboard end, from a line to the salvage ship
—a plate which his immediate predecessor on the job,
Diver Clear, had hooked with a shackle, and which at
one end was now being held up, stretched taut by the
straining wire line to the winch on deck. Getting in
first on all fours under that ton of steel undulating like
a blanket in the water, Blachford next stretched out on
his stomach and wiggled along as far beneath it as he
could get, then thrust the guncotton ahead to the limit
of his extended fingers, jamming it hard between the
lower end of the wobbling plate above and the wreckage on which he lay. Carefully he felt out the lead of
the firing wires to make sure they still ran unbroken
past him to the detonator, and then telephoned up:

"Take in the slack on the firing circuit."

On the surface, a tender hauled in gently on the
electric circuit. At that instant, the wire rope holding
up the plate over Blachford suddenly shot up in loose
bights out of the sea like a broken fishline and fell back
in a tangle of slack coils onto the deck of the diving

ship. Startled, Damant looked at the writhing wire. Something had let go below, the heavy plate it had been holding up had dropped, and Blachford was right under it!

In anguish Damant pressed Blachford's telephone to his lips, feverishly calling his diver. After several very long seconds he got a welcome reply, in strained and measured syllables slowly calling out:

"Give—me—all—the—air—you—can—sir."

Thankful that Blachford, with the crushing load of that steel plate on his back, was still at least alive enough to talk, Damant signaled hastily to raise the air pressure on Blachford's diving line. Immediately the additional air started down the hose, came another call from the agonized diver buried beneath the steel:

"That's right! Give me more yet! And get another diver down here quick!"

As for the last request, that was wholly unnecessary, for already the previous diver, Clear, who had just come aboard after his decompression and was then still in his wet suit, partly undressed, was in hot haste having his weights replaced and his helmet screwed back on, while other seamen hurriedly were reeving off a fresh hoisting wire and some new slings to lift that plate again. But at the request for still more air, Commander Damant paused. The pressure gauge on Blachford's air line already showed a huge excess over what he needed to balance the water at the depth at which he lay; unquestionably under that pressure his suit must be completely ballooned out, and to increase the pressure further meant grave danger of bursting the canvas suit and drowning Blachford immediately. On the other hand, his suit might already be torn somewhere and partly flooded, so that he badly needed the extra air to hold the water from his face as he lay there unable to move under the crushing load of that steel plate on his back.

Should the air pressure be increased or not? Damant was in a terrible dilemma, with Blachford's very life depending on his decision and no help from the telephone in resolving it, for the air roaring now through Blachford's helmet all but drowned out the diver's voice, and very evidently he could not hear Damant's. When carefully the air was throttled down a little to improve the hearing, before Damant ever could get in a word, he heard over and over, slowly articulated by the trapped diver, the anguished plea:

"Give—me—more—air!"

But balancing all the risks, it seemed to Damant unwise to raise the pressure any further, and thus matters stood when Clear, on whom everything now depended, was dropped overboard with the new hoisting sling, to slide directly down Blachford's air hose as a quick guide to the spot where he lay trapped.

Clear landed 20 fathoms down, in the crater of wreckage already blasted through the *Laurentic*, and followed the air hose through the dark water to where it disappeared beneath a twisted steel plate with some loose wire tangled about one end. There was no sign of Blachford save a mass of air bubbles rising in fine clusters from all about that sheet of steel.

Hastily Clear dragged up the fresh wire sling held

by a marline lanyard to his wrist, carefully slipped the clamps of a new wire bridle over the edges of the steel near (but not too near) the free end of the plate. Swiftly but gently he secured the clamps, trying not to jar the plate, working all the time with the knowledge that, aside from Blachford, jammed in under the other end of that sheet of steel was a fulminate detonator buried in guncotton, and that sometimes even more stable explosives than fulminate did queer things. Should his jarring of the wreckage set off that cap, he as well as the trapped Blachford would be blown to bits.

Clear finished securing the new sling, stepped back a little, and shouted into his telephone:

"On deck! Heave round!"

The wire line in the water above him stretched taut, the end of the plate lifted slowly and evenly, exposing Blachford's feet, then his body, soon his helmet. Queerly, in that topsy-turvy world of water, as the distorted steel sheet rose up, Blachford, still nearly horizontal, rose with it, pressing against its under side as if glued there, for with his rig bulging like an overstuffed sausage, he had tremendous buoyancy and could not stay down.

Now a new danger entered. Should Blachford slide out from beneath the steel with that inflated rig, he would "blow up" instantly. And Blachford was helpless to do anything himself to prevent it, for spreadeagled as he was, he could not get his fingers onto his control valves. But Clear could. Seizing his helpless shipmate by one bulging leg to hold him beneath that restraining steel plate lest he float away and suddenly go shooting skyward, Clear reached in, opened the exhaust valve in his helmet wide, and bled the excess air from his suit till Blachford shrank to more normal proportions. Becoming heavy once more, he dropped away from the overhanging plate and thankfully crawled free of it after having been a prisoner for nine minutes.

Fortunately Blachford's copper helmet had protected his head when the plate dropped, or his skull would unquestionably have been crushed by the blow. As it was, the load pressing on him was in a fair way momentarily to break his back, and only the excess air expanding his suit like a pneumatic tire had taken weight enough off him to make the pain bearable till the plate was lifted. Naturally, all he wanted was all the pressure he could possibly get. Damant, solicitously examining Blachford when finally he came up, was not surprised to learn that it had never occurred to his diver, in imminent danger of being crushed to death, that overmuch pressure would rupture the canvas fabric of his suit and drown him.

For two months the blasting went ahead, as monotonously they tore the *Laurentic* apart, with their major excitement the activities of German submarines in that vicinity. Strangely enough no torpedoes came Damant's way, but his divers had ample reason to know that U-boats were working near, for British minesweepers, steaming by them in pairs with long wire sweeps dragging between their sterns, occasionally exploded freshly planted submarine mines. One such went off two miles away, with a detonation coming through the water that struck a diver like a triphammer, giving him a violent and a dangerous shock. After that, whenever the minesweepers got within five miles of his ship, Damant hastily dragged his divers out of the water; but even so, a detonation six miles off soon gave another unlucky diver a severe jar.

However, nothing could be done about it; Damant dared not still further extend his margin or he would have been forced to quit diving altogether, what with the endless forays of U-boats planting mines and the continuous counter-activities of trawlers sweeping them up.

But after two months of blasting, they suddenly forgot all about U-boats. Miller, with a scent unequaled by any of his mates, ran across the gold again! Delving amidst the rubbish sandwiched in between the now uncovered lower deck wreckage lying to port of the flattened-out strong room, he spotted a yellow ingot! Like a hound on a fresh scent he was off, burrowing through the wreckage for more. His time on the bottom soon ran out; Damant signaled him to come up, but nothing could tear him from the job till, having been down 90 minutes and by strenuous exertions having dug out 10 bars altogether, he was satisfied at last and cleared his lifelines of the wreck, so he could be lifted with $80,000 in gold to accompany him.

Here was welcome news for London! Promptly Commander Damant slid into a diving rig and dropped to the bottom to spend a long hour in the depths in that crater they had blasted in the *Laurentic*, checking, in the crushed and crumpled steel all about him, his directions from the remnants of the strong room and what next to tear away to expose more gold. Finally, with a picture of the wreckage vividly impressed on his memory, he started up, taking only a relatively short decompression on his rise.

On coming to the surface, he was somewhat dismayed to find that Miller, between his overlong stay on the bottom, his exertions there, and the excitement attendant on his rediscovery of the gold, had developed another case of "the bends" and had had to be shoved into the recompression chamber for treatment, where he then was, under air pressure again to overcome his torments. Damant peered through a glass port at him. Miller, inside the little single-chambered tank (which was all they could accommodate on their small diving ship) seemed to be already relieved of his pains and resting comfortably, so Damant left him to direct the next diver where best to look for more gold.

But within an hour after emerging from the sea, Commander Damant himself was no longer concerned over gold. He soon found himself in difficulties with one of the rarer manifestations of compressed air disease—his eyeballs began to diverge radically and he began to see double. Then, to top off his troubles, the air bubbles developing in his forehead began to pain him excruciatingly.

Unfortunately for him the only recompression tank they had on board was already occupied by Miller, who would not be out from under pressure for 40 minutes yet, and unless the pressure in that little single chamber were completely released, no one else could enter. With his eyes diverging more and more each minute and the top of his head feeling as if it were about to explode, Damant could hardly wait 40 minutes. Fortunately for him, Miller, inside the tank looking out of the port and glimpsing the most cock-eyed set of optics he had even seen looking longingly in at him, immediately sensed the situation. Chivalrously he blew down the pressure in the tank at once so that the door could be opened to admit his commander, with the result that his own torments promptly returned to double him up in pains worse than ever.

When the door had slammed tight behind him, Damant opened the air valve and started to raise the pressure once more. With gratification he noted as the air roared in and the pressure rose, compressing the bubbles in his head, that the two entirely distinct figures of Miller he saw before him gradually began to approach each other till finally, at 10 pounds pressure, the two Millers coalesced into one only, while at the same time the tortures in his head magically disappeared. However, he had to run the pressure up to double that amount before Miller himself got any relief.

Unfortunately, as the pressure was thereafter gradually reduced, while Damant's troubles was gone for good, Miller's were not. At 4 pounds pressure, Miller again began to suffer the tortures of the damned and there was nothing for it except to jump the pressure up again till his pains vanished and then to try releasing the air at a slower rate. But nothing helped; each time the pressure dropped "the bends" doubled Miller up again. Finally after six and a half hours in the tank (it being then 1:30 A.M.), the temptations of a warm meal and a bed as compared to the chilly interior of that iron chamber overcame the restraints of reason and the two divers at last took a chance and blew down the few remaining pounds of air so they could emerge. But Miller paid for it, for he was soon back in the tank, to suffer all the rest of the night and most of the next day before finally he could rid his system of the nitrogen he had soaked up in his excited grubbing for gold.

So on through the spring and the summer and the early fall, taking the iron wreckage at the bottom of their crater apart bit by bit, and searching amongst the jumble of smashed furniture, bedding, waterlogged provisions, and wooden paneling thus gradually exposed, the divers recovered 542 gold bars, to a total value of about $4,000,000, all of which went promptly back to Britain's hungry Treasury.

After September, wintry weather hit the salvagers once more. It was now apparent there could never be any quick recovery, and something else had meanwhile occurred to put a new complexion on the whole affair. On April 6, 1917, a few months after the sinking of the *Laurentic*, the United States had entered the World War on the side of the Allies. Within a few months of that time arrangements were completed between London and Washington whereby America undertook to finance Britain's purchases in the United States, with the promise of reimbursement later. Britain no longer had to support the pound sterling in New York; while the war lasted, there was no longer any necessity of shipping another ounce of gold westward; and, so far as its effect on the conduct of the war was concerned, the pressure to recover the *Laurentic's* treasure was completely gone.

As a result of this, when winter set in, the Admiralty withdrew its divers from the *Laurentic*, and, while the war lasted, they never came back. Not that Damant and his divers were given any rest even during that winter—quite the contrary—for during the next fourteen months they were kept busy searching the smashed wrecks of sunken U-boats for codes and other useful wartime information.

In the spring of 1919, having been away eighteen months, Damant and his divers returned, a little easier in their minds than formerly as they resumed the job, for they had under them at last a properly equipped salvage ship, the *Racer*, and their fears of submarine attacks were gone.

The *Laurentic*, to their surprise, had changed but little. Apparently the sea had already battered her so flat that not much more crushing was possible, and the divers took up where they had left off, promptly beginning to find more ingots in the wreckage.

But soon a new danger began to threaten. They had been working in way of what had originally been a well deck aft, with forward of them, rising a sheer two deck heights, the after end of the superstructure carrying the first class cabins, and abaft them a similar superstructure carrying the second class accommodations. Oddly enough while the heavy hull plating had given way and folded up, these lighter superstructures had remained fairly well intact, rising like two islands fore and aft of the crushed hull where the divers worked.

With increasing dread, Damant noted that these two superstructures leaned more and more toward each other as he undermined their foundations by tearing away the hull between, but so long as he was finding gold, he was exceedingly reluctant to take his men away from treasure recovery to dismantle the adjacent hull. So, daily keeping an eye on these threatening leaning towers, he kept on delving in the wreckage between, recovering that season some $2,350,000 more. But as the summer drew along, ingots became scarcer and scarcer, and he could only conclude, as has many another miner before him, that his vein of gold was pinching out. Damant decided that in the collapse of the strong room, the gold must have separated into two parts—one part (which he had apparently retrieved) having shot to port through the ruptured bulkhead, while the major part, some $18,000,000 worth, had disappeared through the torn floor of the strong room and must be somewhere buried in the hold underneath everything. With this deduction he was forced to be content when no more bars could be found and winter came to end his labors for 1919.

1920 and 1921 were heartbreakers. The salvagers returned in the spring of 1920 to find that the winter storms had finally torn away the toppling superstructures, spilling them in a mass of twisted steel into their excavation, and filling in the chinks with mattresses, springs, broken china, smashed chairs and tables, tiles and cement from bathrooms, and every conceivable kind of rubbish. To make a complete job of it, the profile of the wrecked hull had been further flattened out so that sand and stones from the sea floor now swept over the broken sides to settle in amongst the wreckage, there to be pounded by the tidal currents and the surging waves into a compact caked mass of conglomerate for which the metallic remains of broken mattress springs formed excellent reinforcements and binders.

For two years, during which very little gold was recovered to encourage anybody, the divers struggled in the sea to remove this debris. Against the sand and the rubbish covering everything, explosives were worthless. Powerful pumps brought in to suck away the sand were equally inffective; and grabs and clamshell dredging buckets got nowhere, partly because of the short periods during which the *Racer* could be held steady over the wreck, partly because of the obstructions below. Nothing showed any effectiveness against this refractory mixture except streams of water from hose nozzles which the divers used to break up the hard-packed sand, after which they hastily filled bags with what sand they had washed loose, and tore away any more substantial wreckage exposed during the washing.

But it was disheartening work, with storms continually washing in fresh sands so that for months it was questionable whether success would ever be possible. In fine weather, the divers gained, and in bad weather, the sand. The discouraged divers found themselves, as storm succeeded storm, beginning to believe that they would never make way against the overpowering forces of the sea. Fortunately, a few scattered bars of gold turned up now and then to revive the drooping divers when defeat seemed inevitable, and finally Damant's ingenuity saved the day.

After every possible mechanical contrivance had folded up in the face of that mixture burying the *Laurentic's* treasure, and it was evident that the bare hands of men alone were all that was left, Damant pulled the job to success by making a hotly contested competition out of the amount of sand each man could dig out in a 30-minute dive under standardized conditions. For 12 minutes with a hose he could wash sand; for 13 minutes thereafter he could pack the sand he had washed free into a near-by sack; and his last 5 minutes he had left to get his bag of sand over to the hoisting bucket to be weighed on deck when he came up.

Scores were carefully kept on what each man brought up; the ingenuity of the divers in fabricating scoops and scrapers to help their speed in digging was amazing; and for the next twenty days the amount of sand brought up per man daily increased as brains came to the aid of brawn and the competition to dig up sand waxed keener. It soon became evident to all that Balson, the strongest diver on the job, was unbeatable, after which interest somewhat declined; but by that time a high standard had been set, below which no diver's pride in the contents of his bag of sand would allow him to fall, and the salvagers started to gain on the sea.

1920 and 1921 dragged wearily away, with very little gold to encourage anybody, and only mountains of worthless sand and an occasional steel plate blasted loose as signs of progress. Still, the hole in the *Laurentic* was continuously getting deeper, and when in 1921 the winter storms arrived to chase the salvagers off, they had at last exposed the shaft tunnel and some of the inner bottom plating near by, so they knew that they had worked their way completely through the ship's hull from top to bottom, and now had before them mainly the sand-filled hold to search.

1922 saw the *Racer* and her crew back again as soon as spring allowed, full of eagerness now that the shell of the ship was near. The first diver down got the surprise of his life when he landed on the bottom of the crater in the *Laurentic's* hulk actually to *see* bars of gold sticking out of the sand! For once the currents had worked on the side of the divers, and during the winter had washed out some two feet of sand which previously had silted in. With a glad cry, the diver started to pluck golden ingots from the sand; before that day was over, 19 bars had been recovered. "They gave themselves up like lambs," the divers reported to Damant.

From that lucky spot the trail of gold led away through the sand toward the port bilge. Daily scraping and digging unearthed more gold nestling in the sand against the shell of the ship, usually one bar at a time buried in hard-packed grit. But the divers nearly went delirious when they came across one nest of 90 ingots surrounded by the broken remnants of the boxes in which originally they had been packed. $750,000 worth of gold went up that day!

That happened only once. The other bars they had to dig for one at a time, completely uncovering the steel skin of the *Laurentic*, section by section, till some 440 square feet had been scraped. By this time· the divers found themselves out to the turn of the bilge with the shell there covered by overhanging deck plating that had been pressed down to within a foot or two of the outer skin, and somewhere between those two layers lay the rest of the gold, millions and millions of dollars of it yet.

To blast away the overhanging plating meant delay and the removal of a shield which was partly holding out the sand. As long as the divers could in any manner squeeze in beneath the wreckage overhead, Damant determined to keep on as he was. So, stretched out flat, headfirst, the divers snaked themselves beneath the broken deckplates to wiggle along on their stomachs over the shell, searching in the deep corrugations (which the collapse of the hull had pressed in the flat shell plating) for gold bars which might have come to rest there.

The corrugations were, of course, all filled with sand, and the digging was hard. Diver after diver wore away his fingernails grubbing through hard-packed sand in these corrugations feeling for ingots. They might have worn gloves, but inside gloves a man lost his sense of touch and wasted time digging out rocks and broken crockery; whereas with his finger tips a man might

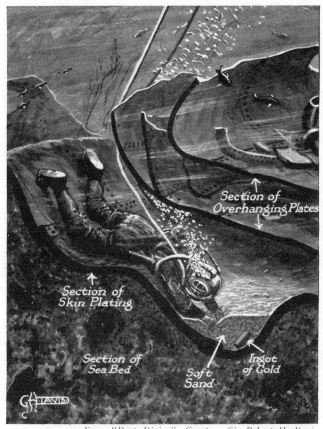

Section of
Overhanging Plates

Section of
Skin Plating

Section of
Sea Bed

Soft
Sand

Ingot
of Gold

From "Deep Diving"; Courtesy Sir Robert H. Davis

GROPING FOR GOLD INGOTS IN THE WRECKAGE OF
THE "LAURENTIC"

soon become expert enough in the feel of gold to pass over without loss of time the rubbish he encountered and save his precious minutes under water for digging out the gold alone.

The divers now had a tough time of it. To avoid having the strong sweep of the current on the long stretch of lifeline and air hose in the water tear them away from their work or perhaps foul the swaying hoses in the wreckage overhead, they tied their lines, when they reached bottom, to a convenient plate, leaving themselves some 30 or 40 feet of slack line to get to their job. Then headfirst they slid in between the waterlogged plates and started burrowing, but it was a ticklish business. As a man's feet were higher than his head, in that position air started to work up from the helmet into the canvas legs of the suit, gradually inflating them so they became lighter and tended to float in spite of heavily weighted shoes. When that happened, there was nothing for a diver to do except on all fours to crawl out backward and stand erect a moment till the water about him had pressed all the air upward again into his helmet, when once more he could slip back into his hole and resume digging for gold with his finger nails.

In this wise, one of the divers, Light by name, was working away trying to dig free an ingot which he could barely feel through the sand with the tips of his outstretched fingers. Tantalizingly his fingers traveled over it, but the refractory bar was so solidly imbedded in the sand it refused to tear free, and so absorbed did Light become in getting a better grip on that $8,000 bar of gold that he completely forgot his own precarious state till suddenly his legs floated upward to touch the overhanging plates, leaving only his helmet bearing on the steel below.

That brought Light back to the realities of his situation, and immediately he attempted to crawl out backward, but with his helmet in the sand and his legs floating up at a fair angle, he no longer was in any position where crawling was a physical possibility.

With his buoyant legs starting to pull him upward, Light hastily took a fresh grip on the ingot (which had gotten him in trouble) to anchor himself down, and shouted desperately into his telephone:

"On deck! Shut off my air!"

On the *Racer*, strange though the request sounded, a tender rushed to comply, but it was too late. Light was light indeed now. Despite his frantic efforts to hang on to that buried gold bar, the pull of his inflated legs was too much for his fingers and tore him loose, to send him shooting up the slope beneath the plating overhead and then, with increasing buoyancy, to float him upside down through the water till he brought up with a jerk on the end of his slack lifeline 40 feet from the bottom, pulling hard against the lashing which held his lines to the wreckage below. There, helmet down, arms stiffly outstretched, feet up, he hung spread-eagled with his rig ballooned out to the uttermost, streaming in the tidal current like an anchored kite balloon.

To make matters worse, Damant, who at Light's first request to shut off his air, had seized his telephone to learn what was wrong, now heard from Light that there was some water in his helmet and he could not tell where it was coming in.

That was bad; every diving suit usually leaks a little, but, except for wetting the diver, the quantity of water entering is of no great moment. However, in Light's case, even a little water was now dangerous. Upside down as he was, a quart of water inside his helmet might well drown him.

It so happened that Blachford, the diver who had preceded Light on the bottom, was still in the water not far below the *Racer's* hull, being decompressed on his way up. Damant promptly had him dragged over by the tenders till he could clutch the air hose running down to Light, then ordered him to slide down that hose to the bottom, cut the lashing that held it there, and ease out on Light's lifeline till he came to the surface.

Down like lead through the sea went Blachford, who, having been once himself in a tight spot on that wreck, knew well enough the need for haste. In less than a minute he was on the bottom, and there at his feet was the lanyard holding his shipmate's air hose, while stretched taut from it, running up to the invisible Light floating somewhere above midway in the depths, was the rest of his air hose.

Blachford drew his diving knife. Getting a good grip with one hand on the hose leading upward to Light, with the other hand he slashed savagely at the lashing and cut it in one stroke. And then things happened fast. In a twinkling, Blachford felt himself shooting upward through the water. So great was Light's buoyancy that, the instant the lanyard was cut, the straining lifeline which Blachford was clutching dragged him surfaceward, with poor Blachford no more able to hold Light down than if he had hold of a stratosphere balloon.

Before the astounded Blachford could let go his grip, he had been jerked so high himself that the air in his own suit expanded enough to spread-eagle him also, and there were both divers, Blachford and Light, with tremendous buoyancy, helplessly racing surfaceward!

A few seconds later in quick succession they shot from the sea and splashed back to float horizontally in their distended rigs on top of the waves, fortunately both of them having missed killing themselves by colliding with the *Racer*. Hurriedly their tenders reeled them in alongside, where Light was hastily taken aboard and jammed immediately into the recompression chamber to get him back under pressure again. Blachford, after enough air had been bled out of his suit to get him vertical once more, was sent down under water to complete his decompression in the normal manner.

Meanwhile the astonished Damant, deeply thankful that neither man had been injured, concluded he had made a mistake. He realized that he should have known that Blachford, none too heavy himself, could never have held Light down. Instead of having him cut the lashing, he decided that he should have ordered Blachford to climb up the line to where Light was floating upside down, grab Light by both feet, and, by making himself as heavy as he could, capsize Light. After

Diver→
Blachford
descending   ←Diver
              Light
              head
              down-
              wards
    Diver Light's
    Air & Life Lines
tied to the Wreck.

From "Deep Diving"; Courtesy Sir Robert H. Davis

DIVER LIGHT "BLOWS UP" FEET FIRST

that, he might have deflated Light's rig enough so that both would have sunk gradually to the bottom again and all might have been well.

However, since both men had come through anyway without mishap, Damant could only chalk his conclusions up as something to be remembered next time, and soon the grubbing after gold bars was going on again as if nothing had happened.

By October, 1922, when stormy weather again stopped the job, the season's work had yielded 895 more gold bars to a total value of $7,500,000 and matters were looking better. Still further to brighten up the picture, 1923 was nearly a repetition of the year before, only better, with 1,255 additional ingots, valued at nearly $10,000,000, recovered from the depths in that one season. By 1924, all except 154 bars, worth about $1,200,000, had been retrieved, and so thoroughly had the divers scavenged the wreckage inside the now exposed shell of the *Laurentic* that it was hopeless to expect to find anything further inside the wreck. Still, $1,200,000 in real gold, the leavings so to speak of six years' work in picking the *Laurentic's* bones, was in itself a substantial prize, more than many a salvage expedition can look forward to as a reward for the whole job.

But where was the rest of that gold? It wasn't inside the shell, so all Damant and his divers could do, when they came back for their seventh year's struggle with the *Laurentic,* was to start digging beneath the shell plating through the rents and openings like portholes in it, into the clay bottom of the sea beneath, in the hope that the missing bars had fallen through the openings and there they might find them.

So down on the shell plating went the divers again, 10 feet now below the level of the adjacent sea floor, fighting to hold back the inflowing flood of sand the while they probed every tear and hole in the steel carpeting the bottom, for ingots beneath. Once in a while they found one, and whenever a diver came to a gap large enough to let him slide through, he started to excavate underneath, with the surprising result that some ingots were discovered under the shell several feet away from the nearest edge of the opening through which they had fallen.

The only possible explanation of this odd phenomenon was that, in the years that had gone by since the gold fell through the holes, the entire shell plating of the wreck must have slipped a few feet, covering the ingots.

Excavating wholesale beneath the shell to recover the lost bars was both slow and dangerous, so Damant decided instead to cut the last layer of the *Laurentic's* bottom up piecemeal, searching the uncovered ocean floor beneath as he went along.

With explosive charges in long strings to do the cutting, the thick steel skin of the *Laurentic* was plate by plate taken apart and the sea floor beneath carefully combed, till an area of over 2,000 square feet had been thus searched. The result of months of this was that 129 out of the last 154 bars had been found buried in the clay when winter finally came to end the search in 1924, leaving still missing only 25 bars out of the 3,211 which eight years before had gone down in the *Laurentic*.

Where to look for those last 25 bars no one could tell, but with 99.2% of all the gold recovered, it appeared that searching another year for the few remaining ingots would not repay the cost of holding back the sand. Accordingly in the autumn of 1924 the job was concluded, eight years after it had started and with seven years actually spent in salvage. The total cost of the work was between 2 and 3% of the bullion recovered.

The recovery of that $24,800,000 in gold from the *Laurentic* stands unique in salvage annals. Over 5,000 dives were made and a large ocean liner was cut to pieces by divers from top to bottom under the worst working conditions imaginable. Work has been done before and since at greater depths, but no submarine job approaching in magnitude and recovery the *Laurentic* salvage has ever been attempted. That no diver was killed nor any even permanently injured, is a tribute to the technical skill with which Damant managed his job, and that such a heartbreaking task was ever successful in the face of practically insuperable obstacles is due only to the extraordinary leadership both on deck and under the sea exhibited by Commander (now Captain) G. C. C. Damant, R.N.

CHAPTER XXII

REALLY deep diving has been handicapped heretofore by the inability of men to do much effective work at depths greater than 200 feet. Until recently, there has been no answer to the problem of *working* at depths greater than that. Armored diving rigs proved no solution, for, while protecting a man from pressure, they prevented him from working, from moving effectively, or from doing anything except acting as an observer for surface operations with grabs, hooks and explosives.

With armored shells for observation it is possible, in deep water, gradually and with difficulty and great expense to tear a wreck to pieces from the surface and to fish up what is inside her, but the divers themselves cannot effectively work on her.

With flexible diving rigs which permit a man to enter a wreck and even under extraordinarily difficult conditions to work inside it, the limit is set to depth by the pressure a man can stand and still retain a reasonable control of his mental faculties, and by the problems of decompressing him on his way up.

Since most naval diving has as its primary object exploring a wreck for the purpose of salvaging the vessel itself or for the recovery of documents of possible military value from a sunken enemy vessel, emphasis is necessarily put on means which allow a diver to work in deep water, rather than on means which enable him merely to observe.

The British have tackled the problem by concentrating on improving the decompression conditions. Their method, devised by Sir Robert Davis and his associates, is to lower a special decompression chamber some 60 feet below the surface, into which the diver is taken at that depth below the water, and then to haul the diver, inside the sealed-up chamber, onto the deck, meanwhile having him finish his decompression while breathing pure oxygen, which greatly accelerates the escape of nitrogen from his blood and correspondingly decreases the period during which he must stay under pressure to escape "the bends."

As an example, using the Davis method, a diver may work 20 minutes at a depth of 300 feet, come up in a total of 13 minutes' time to 60 feet where he enters the Davis chamber, and then be hauled aboard where for the next 73 minutes he is decompressed while breathing oxygen only. After a total decompression time of 86 minutes, he is through and can come out on deck again.

This is a considerable advance. It was not unusual on the S-51 thirteen years ago to give a man that much decompression for a dive in 135 feet lasting not much over an hour, and his decompression was usually spent wholly submerged in cold water, which was a greater strain on a man ordinarily than his hour on the bottom.

A second major difficulty in diving below 200 feet previously has been the difficulty of retaining even a moderate coördination of mental faculties under the heavy air pressure. So far as this may be due to excessive carbon dioxide in the air breathed, the British have improved the situation by adding a cartridge of carbon dioxide absorbent (similar to that used in the "lung") to their diving rig.

As a consequence of these developments, the British now feel warranted in diving to 300 feet, with the confident expectation of working effectively at such depths.

American naval practice has followed a different path toward deep diving. It was recognized long ago that nitrogen troubles were the cause of most diving casualties, and an attempt was made as far back as 1924 to improve the situation by substituting in some degree another gas for nitrogen in the air breathed by the diver.

Working in conjunction with the Bureau of Mines, Navy personnel began experimenting with helium, which the Bureau of Mines had suggested as a desirable substitute for nitrogen.

Partly because of interruptions due to salvage operations on the *S-51* and the *S-4*, and the time spent in "lung" development thereafter, no really worthwhile progress was achieved until recent years, when the work was put in charge of Lieutenant Commander C. B. Momsen who had, after the *S-4* disaster, so prominently figured in making the "lung" a reality in our Navy.

Progress in the use of helium for deep diving has since then been rapid. Momsen, assisted by Surgeons Yarbrough and Behnke of the Navy Medical Corps, working with a crew of naval divers and the facilities of the experimental diving unit at Washington, swiftly developed data enough to make diving with helium mixtures a practicality. Using an atmosphere where helium was wholly substituted for nitrogen, experimental work was carried out on men diving in a tank under any pressure desired, with the medical officers checking the effects on the divers and measuring the relative amounts of helium that were soaked up by a man's blood as compared to nitrogen under similar conditions.

Deductions from theoretical considerations proved to be correct. Helium was found considerably superior to nitrogen. As a gas, its solubility in blood is less than that of nitrogen, and its lightness (helium is roughly only one-seventh as heavy as nitrogen) tends to make it escape from solution faster. Most important of all, it was discovered that under a heavy pressure of helium-oxygen a diver retained a clarity of mind which went far beyond anything previously experienced with natural air at the same pressure.

Another beneficial result of helium, shown by experiment, was that a man could be decompressed safely in less time than with ordinary air.

Only one mechanical drawback to helium was discovered. It is so superior a conductor of heat, as compared with nitrogen, that a body immersed in helium loses heat much faster than in air. Consequently a diver, especially in cold water, may become intolerably cold when using helium in his suit. This was soon rectified, however, by providing the diver with an electrically heated suit of underwear.

With all the data worked up on decompression, experiments were undertaken to verify the use of helium under heavy pressure. Two divers, William Badders (who had worked with me on the *S-51*) and J. H. Mac-Donald were submerged in the diving tank at Washington and put under a pressure equal to a depth of 500 feet of sea water. This was by far the greatest pressure that any man had ever been subjected to. Since there were no untoward effects in decompressing either diver, the scene of operations was moved to the New England coast off Portsmouth, with the *Falcon* as the base ship for diving.

There, with the temperature of the sea floor at 35° F., Diver Badders was dropped overboard from the *Falcon* and went down to a depth of 402 feet. In spite of the cold water, his electrically heated underwear kept him comfortably warm, and his perceptions remained normally acute, in marked contrast to the feeling of extreme "dopiness" felt by divers working with ordinary air at even half the depth in which he was.

Badders came up, decompressed normally, and suffered no ill effects from a dive deeper than any previously made in the open sea.

I believe that this work by our Navy has opened a new chapter in deep sea diving. It should be possible now for trained divers with competent supervisors and properly designed equipment actually to *work* at depths down to 400 feet and even perhaps as far down as 1,000 feet should there be real occasion. There is not at the present time any wreck at such a depth (or even at

lesser depths) worth bothering with, but should the
time come when such wrecks exist, better methods for
working than have existed in the past will be available,
and we may look for men under the sea to work there
more effectively than ever before.

CHAPTER XXIII

ALONGSIDE the dock at the Naval Operating Base in Hampton Roads in late May, 1939, lay the aircraft carrier *Ranger,* awaiting overhaul at the nearby Norfolk Navy Yard. A few days before, far out at sea she had launched from her flight deck all her planes—fighters, scouts, and bombers—to circle like swarms of wasps above her in the air and then straighten away for the invisible land.

Emptied now of her deadly brood, the *Ranger* was moored to the pier, awaiting orders to move on to the Navy Yard where during the coming three months whatever items in hull or machinery needed repair, would receive it.

Down below the waterline in the *Ranger's* hull, I was jammed inside an evaporator shell, watching two seamen dismantling a separator overhead in that confined cylinder. Evaporators, ever since eighteen years before, when first in the Boston Navy Yard, I had designed some for vacuum distillation of salt water, had been one of my pet hobbies. Assigned now to the *Ranger* for a cruise as a reserve officer, with no planes aboard to absorb attention, I was investigating the operation of the *Ranger's* evaporating plant. It was, from the records, doing a good job, but with a few alterations it might perhaps do even better, and with the

chief engineer's permission, there I was dripping perspiration inside that hot shell curiously examining the
baffles intended to keep salt spray from contaminating
the distillate vapors.

A machinist's mate thrust his head through the open
manhole into the evaporator, spotted me coiled up inside.

"Say, commander," he called, "the officer of the deck
wants to see you! They've been passin' the word for
you all over the ship for an hour!"

With some difficulty, I crawled out the manhole,
climbed the steep ladders out of the machinery spaces
to the deck, clad only in soaked dungarees. Near the
starboard gangway was the officer of the deck, scin-
tillating in spotless white. I walked over to him, saluted.

"You want me, sir?" I asked dubiously.

"Yes, commander. Here's something'll interest you!
The submarine *Squalus* has just sunk!"

I looked at him skeptically. The *Squalus?* Submarines were named after fish, but what kind of fish in
Heaven's name was a squalus? It sounded like a joke.

"Quit kidding," I advised. "There isn't any *Squalus*,
fish or sub!"

"There is!" he stated earnestly. "A brand-new sub
just commissioned named that, and she's sunk! Look!"
And he thrust a radio message at me.

Still incredulous, I looked, read,

U.S.S. RANGER, NAVAL DISPATCH:

SUBMARINE SQUALUS DOWN FIVE MILES SOUTH WHITE
ISLAND OFF PORTSMOUTH. UNABLE TO RISE.

It was so. After twelve years of freedom from disaster, another of our submarines was on the bottom!

"They'll probably want you on this," said the officer of the deck. "I've been looking for you for an hour to tell you."

"They'll probably not," I answered. "They've been training officers for this ever since the *S-4*, and they've got plenty now still in active service, but anyway, I'll see. Where's a long distance telephone to Washington?"

At 3 P.M. I was on the wire to the Navy Department, offering my services for rescue work should they be necessary. At the other end of that wire, the Assistant Chief of Naval Operations thanked me warmly, advised me that several hours before by naval patrol plane from Washington, Lieutenant Commander Momsen and the deep sea divers there had already left for Portsmouth, and that the situation seemed covered, but he would get in touch with his chief, and if more help was necessary, I would be notified.

I knew Momsen. I knew his divers. If they couldn't do that rescue job, then it couldn't be done. Satisfied, I hung up.

The afternoon dragged along. I expected nothing, but watched the meager radio reports as they drifted in. Eagerly the flight officer of the *Ranger* offered to fly me up to Portsmouth should I be sent, using one of the two solitary amphibians left aboard the ship. Night came, no word. Regretfully the flight officer informed me that now I would have to wait for morning, that their amphibians were not equipped with instruments to make a long night flight particularly safe; except in dire emergency 'it was unwise to undertake it. I advised him not to worry; the job was covered, I wasn't going.

Then at about 10 P.M., I got orders by radio detaching me from the *Ranger* if I so wished, with permission which came indirectly by telephone, to go to Portsmouth and report to the commandant there in case my services might be used. I left immediately.

It was too late then to catch a train or any transportation out of Norfolk. Night flying from the *Ranger* was also out. I went over the road via Richmond to Washington during the night. In Washington in the very early morning I just managed to catch an Eastern Airlines plane from Mexico headed for Newark Airport, eagerly devouring on the way what little news there was on the *Squalus*. I knew now she was down in deep water, 243 feet; that apparently she was half-flooded due to an open valve; and that 59 men were aboard, of whom 33 were still alive. The others—?

At Newark Airport, I was stuck. All the regular planes going north to Boston were sold out; I could not get aboard one. In this situation, Captain Rickenbacker, head of Eastern Airlines, generously came to my rescue. Eastern Airlines did not fly north of Newark, but as an emergency measure, in case I could possibly help on the *Squalus*, he offered to have me flown direct to Portsmouth in a special transport plane, and if the field there permitted, to land me in Portsmouth. Gratefully I accepted.

In a few minutes, carrying a special meteorologist to keep track of the weather, we lifted smoothly off the ground at Newark Airport, and I was on my way for the last leg. I may say here that never before had I been up in aircraft, nor had I any intention of ever going up. I don't mind diving, but when it came to getting off the ground, I preferred to take my chances on the bottom of the sea rather than up in the air. However, the desire of getting to Portsmouth in a hurry took me aloft, and that morning with Eastern Airlines washed out my previous fears. Both in their regular transport and in their special flight, never anywhere have I seen more efficiently operated and carefully controlled equipment on land or sea.

We made high speed north. Nothing but clouds below us as we cleared New York, but approaching Massachusetts, the clouds faded away, and as we swept over Boston the skies were clear. Passing over Cape Ann, we sighted far ahead the salvage ships off the Isles of Shoals, and a few minutes later we were roaring out to sea headed for the *Falcon*, which a thousand feet below us was moored over the *Squalus*. How far, I wondered, had she got? And what was the situation with the crew of that sunken submarine? I swiftly learned.

The U.S.S. *Squalus*, newest and most modern of our submarines, on the early morning of May 23, 1939, shoved off from the Navy Yard, Portsmouth, New Hampshire, for a routine test dive. The *Squalus*, designed by the Navy and built by the Navy Yard at Portsmouth, was in charge of Lieutenant Oliver Naquin, a submarine officer of long experience, already designated for promotion to lieutenant commander. For some weeks, Naquin and his crew had been running tests at sea, some of which included diving the boat, and eighteen times before this day, the *Squalus* had successfully been dived.

On this day, the specific test to be carried out was a shallow dive from surface condition to fifty feet depth to check the performance of the submarine in its most important military characteristic—its ability swiftly to submerge to periscope depth and disappear from sight, making herself invisible to any rapidly approaching enemy such as a destroyer coming up at thirty-five knots, and at the same time getting a blanket of some ten feet of water over the highest part of her hull (the conning tower and bridge) under which circumstances in wartime shells fired at her would ricochet harmlessly over her across the waves, unable to penetrate that far below the surface.

Out between the low-lying hills at the mouth of the Piscataqua steamed the *Squalus*, running on the surface, driving steadily ahead on her Diesels, steering for deep water to the southward of the Isles of Shoals. Her dive was to be brief, well under an hour. For the depth to which the boat was to be dived, deep water was not required, but long experience in operating submarines, especially new ones, had demonstrated that with vessels as long as the *Squalus,* about 300 feet from stem to stern, it is highly desirable to have plenty of depth under her so that should she take an undue trim while planing under, her crew would have sufficient water under her keel to regain control of their lengthy vessel before either stem or stern crashed bottom.

Aboard the *Squalus*, aside from her commander, were four officers, fifty-one seamen of various ratings, and three civilians, one the test engineer of the Navy Yard and the other two representing contractors who had supplied machinery.

Some five miles to the southward of the Isles of Shoals and about the same distance off the coast of New Hampshire, with the charts showing 243 feet to the bottom, the *Squalus* at 8:30 A.M. prepared to dive. Below at their diving stations were the crew—torpedo men and officers in the forward and after torpedo rooms at bow and stern; machinist's mates at the Diesels; electricians standing by motors, storage batteries, and switchboards ready for submerged operation; and in the crowded control room the diving officer and his enlisted assistants. Before these latter to port was a maze of diving wheels, depth gauges, valve-operating gear, indicating lights of various colors, and a galaxy of gauges with which to flood the boat down to proper depth and then maintain her there at proper trim. Alongside them, poised, watch in hand, waiting to check the time from full ahead on the Diesels to level trim at 50 feet depth driving submerged on her electric motors, was Preble, the Navy Yard's civilian test engineer.

The day was clear but the surface fairly rough, when Lieutenant Naquin on the topside gave the order to dive. In every compartment of the boat below, the diving signal shrieked. On the bank of indicators in the control room before Alfred G. Prien, machinist's mate, second class, twenty-six years old with six years of naval service, whose job it was to flood down for sub-

mergence, the diving light flashed on. Alongside him
Preble started his watch, as Prien swung into action.
It was Prien's task to open the vent valves through
which the air in the ballast tanks escapes while they
fill with water; to open the kingston valves at the bot-
toms of these same ballast tanks to admit the water
which floods them and sinks the boat; and most important of all to close the two huge main air induction
valves high up in the submarine's superstructure abaft
the conning tower, through which valves while on the
surface air enters to feed the Diesels and to ventilate the
hull compartments.

Prien noted that the Diesels were stopped, thus giving him clearance to close the outside air induction
valves. Immediately he thrust over to "Closed" the
control gear for the air inductions. Through high pressure lines, liquid flowed from his controllers to the
hydraulic operating cylinders at the valves themselves
to push them down to the closed position. Swiftly on
the bank of electric indicators before him, the red lights
showing the valves open went out; in a few seconds,
the green lights showing them closed, flashed on. That
done, reported to the diving officer, Lieutenant Doyle,
and checked by him, Prien proceeded to his next operation, shoving over to "Open" the controllers for the
vent valves and the flood valves scattered out of his
sight over a length of two hundred feet of the submarine's hull.

From the ballast tanks, air roared out the vents on
top as the sea rushed in through the opened flood valves,
and on the depth gauges the needles started up the

dials as the boat flooded down. At the diving wheels, the diving rudders were trimmed downward to plane the boat under. At the switchboard, the main motors were started up, and running now on power from her storage batteries, the *Squalus* planed smoothly downward, the throbbing of her Diesels missing now inside the silent hull.

At a little beyond twenty-five feet on the depth gauge dials the *Squalus*, with buoyancy practically all gone, planed her deck beneath the surface. Nearing thirty-five feet, the tops of the two main air inductions, one 31 inches in diameter, the other about 18 inches wide, both shielded by the conning tower forward of them from the rising water, went under, and about this time, Lieutenant Naquin, his surface navigation finished, descended to the control room.

Forty feet on the depth gauges and the last vestiges of the *Squalus'* bridge and conning tower disappeared beneath the sea, leaving only an agitated wake, boiling with the air ejected from her ballast tank vents to mark where she had been. Still driving ahead on her electric motors the submarine planed downward the last few feet and as the depth gauge needles approached 50, the diving officer started to level off the boat with the horizontal rudders, and Lieutenant Naquin started for his station at the periscopes.

Inside the *Squalus*, everything seemed right—air induction indicator lights showing both valves closed, air pressure in the boat normal, the dive in every respect similar to the eighteen uneventful plunges the submarine had previously experienced. Alongside the control men, Preble, checking the diving time, looked tentatively at his watch, nodded approvingly, prepared to take the time at the instant the trim indicators showed "Level." It would be a good dive.

And then, startling in its suddenness, came disaster. From the engine room, over the telephone into the control room, came the almost unbelievable message, the first and last that ever came from aft,

"WATER COMING INTO THE ENGINE ROOM!"

As that message echoed through the control room, Prien, dumbfounded, stared at his indicator lights. They showed "Closed."

Inside the *Squalus*, things began to happen fast. The air inside the boat, compressed by the water rushing in aft, rose suddenly in pressure, painfully distending ear drums and instantly warning every man in the boat that something was wrong. On the heels of that, under the impact of the sea flooding in aft, down trimmed the stern, giving a sharp "up" angle to the deck.

What was wrong? Undoubtedly the eyes of both captain and diving officer instinctively flew to the main air induction valve controls and indicators. Had either showed "Open," down on those controllers would have leaped an officer to close the overlooked valve, and by that closing to shut off instantly the deadly stream of incoming water. But nothing like that happened. On valve gear already in proper closed position and with indicator light showing "Closed," nothing more can be done, and Naquin wasted no further time on that. The trim by the stern was increasing, on the depth gauges the needles were starting to race up the dials from "50," the boat was going down. The sole remaining emergency measure was taken.

"BLOW ALL BALLASTS!" shouted Lieutenant Doyle.

On the compressed air manifolds, tense operators swung open the valves, admitting through reducers high pressure compressed air from the ship's banks to the ballast tanks in a final attempt to expel the water recently admitted to the ballast tanks, lighten up the boat, and if possible float her up to the surface.

But water was pouring in aft too fast for that. The trim rapidly increased, reaching at least a thirty degree angle, at which slope, except by clutching something fixed, no one could stand on deck. Prien, unable yet to understand what was happening, slid backward from his post, saved himself by grabbing a control lever, clung there, eyes staring at the indicator lights of his valves, still showing "Closed." Forward, as yet free of all water, doors began to close, internal valves to swing shut, as the torpedomen started to seal off. In the after battery room, just abaft the control room, filled with the crew's bunks, water started to gush in from overhead through ventilation valves and blowers. Five men stationed there immediately made a dash for the open door to the control room. Booth, radio man, testing storage batteries beneath the deck, racing up the steep slope made it first, crawled through into the control room without much difficulty, to find there at the bulkhead Maness, electrician's mate, third class, reaching through from the control room to close the heavy steel door.

"Hold it open!" shrieked Booth. Maness paused.

Booth's shipmates in the after battery room, some dozen of them, were less fortunate. Between water gushing down the deck and the rapidly increasing angle, climbing became next to impossible. Four more men only got through, of whom the last, O'Hara, was several times washed back by the flood before he at last fought his way up to the door and crawled through, the last man to escape from aft. And as he, half drowned, came through the door, Maness, by a superhuman effort, managed to drag the heavy mass of steel forming that door upward against the slope, and slammed it shut, immediately setting up the dogs to hold it watertight.

Meanwhile in the control room, in spite of ballast water going overboard, the depth gauges showed increasing depth, the angle of trim increased, and stern first, the *Squalus* plunged to the bottom. Within three minutes she hit, slowly leveled off as her bow dropped, and came to rest with the bow trimmed up at an angle of eleven degrees.

In another moment, the ship's power went off and the lights flickered out, excepting only a few on an emergency lighting circuit, which in perhaps two minutes more short circuited also and left the ship in darkness.

Coolly, calmly, Lieutenant Naquin went about securing the unflooded section of his boat. In the control room, the forward battery room, and the forward torpedo room, all interior valves were closed, compartment doors closed, and every available flashlight in the boat was hastily gathered together to be used only under the captain's directions.

On a swift telephone check, Naquin learned the men in his torpedo room forward were safe, as were a few also in the forward battery room between him and that torpedo room. But from no one aft was he able to get the slightest response by telephone or by hammering. Realizing then that those aft were beyond his help and that inside a sunken, dark, and half-flooded submarine his major problem was to save those still alive, he prohibited all discussion or reference to those astern as the first step in keeping up morale.

As his second step, once everything had been checked for absence of leaks, the few men in the battery room were brought aft into the control room to save them from possible chlorine gas which might there be generated if any salt water leaking in got to the sulphuric acid inside the lead cells, or on any acid which might have spilled while the boat, at a terrific angle, went down. After that, no man except on the captain's orders, was permitted to move about. Instead, wrapped in such blankets as could be secured from the officers' quarters in the battery room, the mattresses dragged from the same spot, and the little bedding available from the crew's bunks in the torpedo room, each was ordered to lie quietly and avoid any exertion which by accelerating his breathing would increase the amount of carbon dioxide discharged from his lungs and thereby hasten the fouling of the air.

What little else Naquin could do was soon done. Over the telephone, he ordered Lieutenant Nichols in

the torpedo room to let go from inside there the releasing gear on the *Squalus'* marker buoy, and in a moment up from the bow of the helpless submarine to
the surface rose that buoy carrying in its cable a telephone line, and inside the buoy itself, marked by an
engraved plate with the words,

"SUBMARINE SQUALUS SUNK HERE. TELEPHONE IN-
SIDE. BOW BUOY."

a telephone sealed watertight.

And then the trapped men concentrated wholly on
keeping life in their freezing forms. For outside on the
bottom of the sea the temperature was thirty-one degrees Fahrenheit; inside the *Squalus*, with no heat available except what their own poor bodies could supply,
the temperature quickly dropped to thirty-six degrees.
There in the darkness, in the silence, in the chilly damp-
ness of their steel prison, the trapped men, alone with
their thoughts, waited for rescue.

Nine-thirty A.M. came and the *Squalus*, which should
be reporting the completion of her dive by radio to
Portsmouth, was unheard from. Shortly after, a sister
submarine, the *Sculpin*, just starting south on a long
cruise, was asked by Portsmouth to look over the area
of the *Squalus'* dive for any sign of her. And when
through the depths the vibrations of her sister's screws
beat down against the steel sides of the *Squalus* in a
rhythm easily recognized there, Naquin immediately
ordered the discharge through an ejector of special
bombs which, on reaching the surface, ignited to make

a decided smoke, and first directed the *Sculpin's* attention to where to look.

Steaming over there, the *Sculpin* soon sighted the marker buoy, took it aboard, quickly learned from Lieutenant Nichols, below in the torpedo room, the plight of the *Squalus*. Naquin himself, coming forward to the torpedo room to give further instructions, had hardly more than exchanged greetings with his fellow captain on the surface, when a sea hit the *Sculpin*, drove her to leeward and parted the buoy cable, stopping all further discussion by phone, and worst of all, leaving the *Squalus* with no surface line or buoy exactly to mark her position and guide divers down to her.

On the *Sculpin*, the radio began flashing in the news. In a few minutes more, in far-off Washington, in the submarine base at New London, at every naval station on the East Coast, diving gear, divers, and salvage gear were being mobilized for transport to Portsmouth. From Washington, Lieutenant Commander C. B. Momsen, inventor of the escape "lung" and for several years officer in charge of the Naval Experimental Diving Unit, together with his divers, were loaded aboard naval patrol planes and started through the air to the rescue of their comrades on the bottom of the sea. From New London, the Navy's Submarine Base, on board the rescue ship *Falcon* sailed the divers there. The *Falcon*, specially equipped for diving, veteran of the salvage campaigns on *S-51* and *S-4*, was almost as important as the divers themselves, for on her decks she bore a special rescue bell, compression chambers for treating the divers to avoid "the bends" after a dive, and all the intricate gear required for handling divers in deep water.

In some ways, the *Squalus* was lucky. Since the days of the *S-4*, our Navy had developed, under Momsen's direction mainly, the technique of diving with a synthetic atmosphere composed wholly of helium and oxygen. With this, instead of natural air, our divers can go much deeper than ever before, and what is more important, instead of the dazed state in which the diver under heavy pressure in the deep sea formerly worked, it permits the diver to keep his faculties as keenly acute as if he were only at a shallow depth. A few years ago, work at 243 feet where the *Squalus* lay would have been extremely dubious if not impossible; now with helium, it was well within practicable diving range.

Next, even if the divers failed, the *Squalus* was fitted with escape chambers and escape "lungs." Wearing one of these, the development of Momsen, himself once a submarine commander, a man might emerge from the submarine and rise unaided through the sea to the surface, floated up by the buoyancy of the artificial "lung" strapped to his chest and meanwhile breathing into it as he rose, through a special mouth connection. This device, to the use of which every submarine man is trained in 100 feet of water, was available as a last resort to Naquin and his 32 companions, but except as a last resort, Naquin was reluctant to resort to it, for the water was far deeper than any "lung" had yet been tested; further and worse, the ship was literally submerged in ice water and whether his men, already nearly frozen, could escape complete numbness once they emerged into the sea for their five minute rise to the surface, was highly questionable. And upon every man's keeping his head during that long rise through the sea his life depended, for a man getting panic-stricken when he found himself under the terrific pressure of the sea, with nothing in sight in any direction but water, and instinctively under those circumstances attempting to hold his breath, would promptly as a result rupture his lungs and seal his death warrant.

Naquin well knew that aboard the *Falcon* was a special rescue bell, the invention of Commander Allan McCann, which might obviate the need of using the escape "lungs." This device, intended to be hauled down after a diver had attached the necessary line on top of one of his submarine hatches, would permit the passage into it through the opened hatch of some eight men, who thereupon, when the lower entrance to the bell was closed, might ride upwards inside it to the surface never at any time exposed to the sea.

For this device, Naquin decided to wait rather than risk his men in the "lungs."

The wait was long. By evening Momsen and the first divers had arrived by air. Also by that time, a grappling hook dragged from the surface had caught in the *Squalus*, making good the guide line lost when the marker buoy line parted. But until the *Falcon* arrived diving was impracticable, and the *Falcon*, steaming full power from New London, was not due till early morning.

All through that endless night, 33 freezing men lay in darkness and in silence at the bottom of the sea. The air grew worse, fouled by carbon dioxide, and the

air-purifying system was started up to remove the gas.
No one talked, no one moved, no light showed in that
Stygian blackness save occasionally the flicker of a
flashlight as some fitting was checked for leakage. With
insufficient blankets for so many men, those available
did more than double duty to wrap the shivering seamen. Nothing disturbed the silence except the occasional rapping out with a hammer in Morse code of
some message sent by Naquin to the ships on the
surface.

Morning came, the *Falcon* came, and with both came
the finest diving weather that divers have ever been
blessed with. Overhead a brilliant sun to light the
depths, on the surface of the sea only a light swell insufficient even to cause moderate rolling. In strong contrast to that bleak December day in 1927 when a fierce
northwest gale off Provincetown stopped dead this same
Falcon and her divers for three days while below them
on the *S-4* slowly perished from cold and fouled air
the crew trapped below, the conditions on the surface
were now ideal. Hurriedly the *Falcon* ran out her four
moorings, bow and stern, to hold her fixed while the
divers worked below. That done, aboard the *Falcon*
came the grappling line caught somewhere in the
wreck below. And down that line at 10:10 A.M. slid
Martin Sibitsky, bosun's mate, first class, diver from
the Submarine Base at New London, clad in electrically
heated underwear to keep him from freezing to death,
enclosed in a ponderous flexible canvas-and-rubber diving suit weighted down with 200 pounds of lead and
copper.

Swiftly Sibitsky dropped down the line to the bottom, to his great joy found that the grappling hook had caught so that he was able to land on the deck of the *Squalus* close to the forward torpedo room. Favored further by a most remarkable visibility, unusual for northern waters, of at least sixty feet, he resecured the grappling line, then plodded along the deck forward through the sea, dragging with him one end of the hauling-down wire for the rescue bell.

He came to the hatch over the torpedo room, knelt on the motionless deck, carefully shackled to a spindle in the center of the hatch the wire he carried. And then Sibitsky's work was done. Pushing slowly aft, he came to the guide line, signaled to be hauled up fifteen minutes after he hit bottom, and started his slow rise to the surface for decompression. At 11:24 A.M., a little over an hour after going overboard, he was once again aboard the *Falcon*.

Overboard from the *Falcon* went the rescue chamber, to some degree a miniature submarine in its functions, its two operators, divers Badders and Mihalowski, inside to work its machinery and ballast tanks. Balanced off till it was only slightly buoyant, its air motor reeling in the guide wire, down it went through the sea till it came to rest on the *Squalus'* deck over the hatch. Then began the delicate process of attaching the bell watertight to the submarine, a process in which by blowing, venting, and making use of a thick rubber gasket on the bottom of the bell, the operators in it made a vacuum cup out of the bottom of their chamber, as a result of which the sea pressure itself forced the

CHIEF MACHINIST'S MATE WILLIAM BADDERS, U.S.N., IN THE HELIUM-
EQUIPPED DIVING RIG USED ON U.S.S. "SQUALUS"

bell so tightly to the submarine's hull that leakage was impossible.

And at that instant, just as a vast discharge of air bubbling up from the depths showed that the rescue chamber was seating on the *Squalus* and the water between was being expelled, overhead roared the Eastern Airlines special plane to swoop down on the *Falcon* like an eagle on its prey.

Looking down on the *Falcon* from the air as I flew over her and then round and round in heavily banked circles, I watched the sea boil up like a light green fountain beneath her hull. Spread out below me was the whole rescue picture—the *Falcon* with her lines vanishing into the sea going far down to the bottom where her men were attaching the rescue chamber, nearby the *Sculpin* anchored, a little away several tugs, racing up from the south the cruiser *Brooklyn* with more equipment and more divers. But beneath me as from the air I peered down on the *Falcon*, it was evident that the task of rescue was far advanced, that Momsen had that job thoroughly under control, that from no source at all was help then needed—except such as was wholly beyond human power, to insure the continuance of the remarkably fine weather conditions favoring the rescue fleet.

Far down on the bottom of the sea, the rescue chamber, firmly pressed down on the *Squalus*, was ready for the last step. Through the lower compartment of that bell, emptied completely of water, the operators descended to open the exposed hatch to the torpedo room of the *Squalus*. Gathered below the hatch were Naquin and his men. Hurriedly the captain told off those to make the first trip to safety—Lieutenant Nichols to inform the officers on the *Falcon* of conditions inside the *Squalus*, and six seamen chosen from among those whose weakened condition made hasty release imperative if they were to survive.

Up through the submarine's hatch into the bell, dragged to safety by the strong arms of Badders and Mihalowski, went the first seven to be rescued. Immediately the last man had been dragged up, the lower hatch of the bell promptly closed to seal them in. And on the remaining 26 men in the *Squalus* down slammed their own hatch sealing them once again in their prison.

Quickly the vacuum holding the bell down was broken, the wire slacked off, and the bell started up with its cargo of the saved, soon to reach the surface, where to its seven inmates, as they climbed out through the top hatch to the *Falcon*, the sunlight overhead seemed deliriously unbelievable.

Twice more during the afternoon the bell repeated its journey, removing all but nine of the crew. In darkness the last trip was started, the last contact made with the *Squalus*, the last nine men, including Lieutenant Naquin, taken aboard, and for the last time the hatch to the submarine closed, leaving her on the bottom of the deep sea, utterly silent, peopled only by the dead.

Up now from the depths went the rescue bell, only to the horror of those on the *Falcon* to jam its hauling wire at 150 feet from the surface, unable to rise farther!

In the darkness above, the *Falcon* sought to aid with its divers, but there was nothing they could reach to clear the jam. Suspended in the icy sea, inside the bell the operators strove futilely to clear their fouled machinery; outside the divers just as futilely fought the sea to aid. But nothing could be done to clear the mechanism and finally, after three desperate hours below the bell, the divers cut the wire and left the *Falcon* with the delicate job of getting the chamber to the surface with its precious cargo—a task in which those on deck the *Falcon* and inside the bell had to juggle buoyancies and weights so that neither an overheavy bell would break its hoisting line and go crashing to the bottom with parted air lines, leaving its inmates to asphyxiate alongside the *Squalus*, nor an overlight bell go shooting to the surface to smash itself against the *Falcon's* hull.

Near midnight, after four hours on its journey, the bell came up successfully, discharging Naquin and his eight companions and the rescue job on the *Squalus* was complete—the first time in history that from any such depth men have been saved from a foundered submarine. And the labors of the Navy over the twelve years since the *S-4* disaster in 1927 to train more divers, to provide better rescue methods, and to work successfully in deep water had at last borne fruit. Thirty-three men taken from the *Squalus*, who under former conditions would at 243 feet down surely have been beyond human aid, are a living proof that those whose deaths in the *S-4* started the nation-wide demand for better methods of saving submarine crews, did not die in vain.